Analytic Combinatorics

A Multidimensional Approach

Discrete Mathematics and Its Applications

Series Editors
Miklos Bona
Donald L. Kreher
Douglas West
Patrice Ossona de Mendez

Advanced Number Theory with Applications
Richard A. Mollin

A Multidisciplinary Introduction to Information Security
Stig F. Mjølsnes

Combinatorics of Compositions and Words
Silvia Heubach and Toufik Mansour

Handbook of Linear Algebra, Second Edition
Leslie Hogben

Combinatorics, Second Edition
Nicholas A. Loehr

Handbook of Discrete and Computational Geometry, Third Edition
C. Toth, Jacob E. Goodman, and Joseph O'Rourke

Handbook of Discrete and Combinatorial Mathematics, Second Edition
Kenneth H. Rosen

Crossing Numbers of Graphs
Marcus Schaefer

Graph Searching Games and Probabilistic Methods
Anthony Bonato and Paweł Prałat

Handbook of Geometric Constraint Systems Principles
Meera Sitharam, Audrey St. John, and Jessica Sidman

Additive Combinatorics
Béla Bajnok

Algorithmics of Nonuniformity: Tools and Paradigms
Micha Hofri and Hosam Mahmoud

Extremal Finite Set Theory
Daniel Gerbner and Balazs Patkos

Analytic Combinatorics: A Multidimensional Approach
Marni Mishna

https://www.crcpress.com/Discrete-Mathematics-and-Its-Applications/book-series/CHDISMTHAPP?page=1&order=dtitle&size=12&view=list&status=published,forthcoming

Analytic Combinatorics

A Multidimensional Approach

Marni Mishna

CRC Press
Taylor & Francis Group
Boca Raton London New York

CRC Press is an imprint of the
Taylor & Francis Group, an **informa** business

A CHAPMAN & HALL BOOK

CRC Press
Taylor & Francis Group
6000 Broken Sound Parkway NW, Suite 300
Boca Raton, FL 33487-2742

First issued in paperback 2022

© 2020 by Taylor & Francis Group, LLC
CRC Press is an imprint of Taylor & Francis Group, an Informa business

No claim to original U.S. Government works

ISBN 13: 978-1-03-247513-4 (pbk)
ISBN 13: 978-1-138-48976-9 (hbk)
ISBN 13: 978-1-351-03682-5 (ebk)

DOI: 10.1201/9781351036825

Visit the Taylor & Francis Web site at
http://www.taylorandfrancis.com

and the CRC Press Web site at
http://www.crcpress.com

To Madeleine, Eleanor, Felix and Cedric.

Contents

Preface   xiii

Symbols   xvii

Welcome to Analytic Combinatorics   xix

I Enumerative Combinatorics   1

1 A Primer on Combinatorial Calculus   3
  1.1 Combinatorial Classes   4
  1.2 Words and Walks 4
    1.2.1 Words and Languages 5
    1.2.2 Lattice Walks 6
    1.2.3 What Is a Good Combinatorial Formula? 7
    1.2.4 Bijections 7
    1.2.5 Combinatorial Operations 8
  1.3 Formal Power Series 11
    1.3.1 Ordinary Generating Functions 12
    1.3.2 Coefficient Extraction Techniques 13
  1.4 Basic Building Blocks 14
    1.4.1 Epsilon Class 14
    1.4.2 Atomic Class 14
    1.4.3 Admissible Operators and Generating
        Functions 15
  1.5 Combinatorial Specifications 17
  1.6 S-regular Classes and Regular Languages 18
    1.6.1 Finite Automata 19
  1.7 Tree Classes 21
    1.7.1 Lagrange Inversion 23
  1.8 Algebraic Classes 24
  1.9 Discussion 28
  1.10 Problems 29

2   Combinatorial Parameters   **35**
  2.1   Combinatorial Parameters   36
      2.1.1   Bivariate Generating Functions   36
  2.2   What Can We Do with a Bivariate Generating
      Function?   37
      2.2.1   Higher Moments   38
      2.2.2   Moment Inequalities and Concentration   40
  2.3   Deriving Multivariate Generating Functions   41
      2.3.1   Multidimensional Parameters   41
      2.3.2   Inherited Parameters   42
      2.3.3   Marking Substructures   43
  2.4   On the Number of Components   46
  2.5   Linear Functions of Parameters   46
  2.6   Pathlength   47
  2.7   Catalytic Parameters and the Kernel Method   49
  2.8   Discussion   51
  2.9   Problems   51

3   Derived and Transcendental Classes   **55**
  3.1   The Diagonal of a Multivariable Series   56
      3.1.1   The Ring of Formal Laurent Series   58
      3.1.2   Basic Manipulations   59
      3.1.3   Algebraic Functions Are Diagonals   60
      3.1.4   Excursion Generating Functions   61
  3.2   The Reflection Principle   64
      3.2.1   A One-dimensional Reflection   65
      3.2.2   A Two-dimensional Reflection   67
  3.3   General Finite Reflection Groups   69
      3.3.1   A Root Systems Primer   69
      3.3.2   Enumerating Reflectable Walks   71
      3.3.3   A Non-simple Example: Walks in A_2   72
  3.4   Discussion   74
  3.5   Problems   75

II   Methods for Asymptotic Analysis   **79**

4   Generating Functions as Analytic Objects   **81**
  4.1   Series Expansions   82
      4.1.1   Convergence   82
      4.1.2   Singularities   83

4.2 Poles and Laurent Expansions 84
    4.2.1 Puiseux Expansions 86
4.3 The Exponential Growth of Coefficients 87
4.4 Finding Singularities 91
4.5 Complex Analysis 92
    4.5.1 Primer on Contour Integrals 92
    4.5.2 The Residue of a Function at a Point 93
4.6 Asymptotic Estimates for Meromorphic Functions . . 95
4.7 The Transfer Lemma 98
4.8 A General Process for Coefficient Analysis 99
4.9 Multiple Dominant Singularities 102
4.10 Saddle Point Estimation 106
4.11 Discussion 108
4.12 Problems 108

5 **Parallel Taxonomies** **111**
5.1 Rational Functions 112
5.2 Algebraic Functions 113
5.3 D-finite Functions 115
    5.3.1 Closure Properties 116
    5.3.2 Is It or Isn't It? 117
    5.3.3 G-functions 120
    5.3.4 Combinatorial Classes with D-finite Generating
        Functions 120
5.4 Differentiably Algebraic Functions 121
5.5 Classification Dichotomies 123
5.6 The Classification of Small Step Lattice Path Models . 124
    5.6.1 A Simple Recursion 125
    5.6.2 Models with D-finite Generating Functions . . 127
    5.6.3 Models with Non-D-finite Generating
        Functions 129
5.7 Groups and the Co-growth Problem 130
    5.7.1 Excursions on Cayley Graphs 131
    5.7.2 Amenability vs. D-finiteness 132
5.8 Discussion 133
5.9 Problems 136

6 **Singularities of Multivariable Rational Functions** **139**
6.1 Visualizing Domains of Convergence 140
    6.1.1 The Univariate Case 140
    6.1.2 The Multivariable Case 141

6.2   The Exponential Growth   144
6.3   The Height Function   146
6.4   Visualizing Critical Points   148
6.5   Examples   149
      6.5.1   Delannoy Numbers   149
      6.5.2   Balanced Words   150
6.6   Discussion   152
6.7   Problems   152

7   Integration and Multivariable Coefficient Asymptotics   155
7.1   A Typical Problem   156
7.2   Warm-up: Stirling's Approximation   157
7.3   Fourier-Laplace Integrals   159
7.4   Easy Inventory Problems   160
7.5   Generalizing the Strategy to Higher Dimensions . . .   162
      7.5.1   Multivariate Cauchy Integral Formula   162
      7.5.2   A Formula for Fourier-Laplace Integrals   162
      7.5.3   How Not to Transform This Integral   163
7.6   Example: Simple Walks   164
      7.6.1   Exponential Growth   164
      7.6.2   Estimating Cauchy Integrals   165
7.7   A More General Strategy   167
7.8   Discussion   172
7.9   Problems   174

8   Multiple Points   177
8.1   Algebraic Geometry Basics   178
8.2   Critical Points   180
8.3   Examples   182
      8.3.1   Tandem Walks   182
      8.3.2   Weighted Simple Walks   184
8.4   A Direct Formula for Powers   185
8.5   The Contribution of a Transverse Multiple Point . . .   186
8.6   Discussion   188
8.7   Problems   189

9   Partitions   191
9.1   Integer Partitions   192
9.2   Vector Partitions   194
      9.2.1   Integer Points in Polytopes   196

9.3 Asymptotic Analysis 197
   9.3.1 The Singular Variety and Hyperplane
          Arrangements 198
   9.3.2 Reducing to the Case of Transversal Intersection 199
   9.3.3 Algebraic Independence 201
   9.3.4 Decomposition Dictionary 202
   9.3.5 Decomposition into Circuit-free Denominators 202
   9.3.6 The Complete Reduction Algorithm 203
   9.3.7 How to Compute a Reduction Rule 204
9.4 Asymptotics Theorem 204
9.5 An Example 205
   9.5.1 An Exact Solution 206
   9.5.2 An Asymptotic Solution 207
   9.5.3 The Bases with No Broken Circuits 207
   9.5.4 The Reduction Algorithm 207
   9.5.5 Asymptotic Formula 209
9.6 Discussion 210
9.7 Problems 211

Bibliography   213

Glossary   225

Index   227

Preface

One of the most intriguing mathematical deductions that first captures our attention is Cantor's diagonal argument. It can be used to prove that real numbers cannot be put into one-to-one correspondence with the set of natural numbers $1, 2, 3, \ldots$. The result is something that surprisingly can even be shown. The process creates a pathological object by extracting along a diagonal.

Here we consider sets of discrete objects – combinatorial classes – and try to understand their structure and behaviour. We start with constructions built very naturally, whose enumeration is essentially systematic. They are the foundation of objects and techniques that are more complex. We then build classes of increased complexity by extracting along a diagonal. We are able to give structure to a concept of "transcendental" families of objects.

To illustrate the notions, techniques and eventual taxonomy, we draw many of our examples from formal languages and lattice walks. These two grand themes intersect combinatorics in different but complementary ways.

Power series are the main workhorse here, encoding combinatorial data using mathematical functions. At first they are mere data structures, famously used as "clotheslines for numbers". However, in the second part we consider them as analytic objects, and there is a deeply revealing, and intriguing interplay between combinatorics, geometry, and transcendence that we explore.

Indeed, the world of formal power series offers many echoes of number theory – historically humanity has asked many of the same questions about transcendency, finite representation and computability. These questions are very natural here, particularly when they are closely tied to combinatorial families. As far back as Hadamard in the early 1890s, mathematicians have been fascinated by the relationship between the coefficients of a power series and the properties of the function it represents.

The behaviour of a series at the edge of the domain of its convergence – the singularities and the natural boundaries – are particularly interesting. Our series coefficients are counting sequences, and so the coefficients are all non-negative integers. This gives an important structure that we can leverage to say very nontrivial things about the combinatorial classes we study. We will try to understand this interface and say clever things about combinatorial classes.

The text is intended to introduce the main concepts and intuition of analytic combinatorics so that you might broaden your personal toolkit to study discrete objects. As you might guess from the size, this text is not comprehensive – far from it! It is intended to act as a gateway to more detailed works that can be intimidating in their completeness and mathematical sophistication. We mention two such reference books by name: *Analytic Combinatorics*, by Flajolet and Sedgewick, and *Analytic Combinatorics in Several Variables* by Pemantle and Wilson. The reader intrigued by the ideas and interactions between complex analysis, geometry and combinatorics will be quite delighted to learn that the stories presented here continue in so many deep directions. These topics will lead you through number theory, algebraic geometry, formal language theory and probability.

This book was made possible because of the support of many people, that I acknowledge with a tremendous debt of gratitude.

Thank you Miklos Bonas, who encouraged this text and made possible a connection with the team at CRC press that worked with me. In this breath I must mention the BIRS research station, where conditions were ideal to develop not only this project but the general sharing of ideas.

The naissance of the text was graduate notes, built with Andrew Rechnitzer, and I owe much to him for insisting on a casual tone. Said notes were tested on numerous groups including multiple generations of Simon Fraser University graduate and undergraduate students. Thank you also to Anthony Guttmann for feedback.

The sections on analytic aspects of diagonals was developed for a series of advanced minicourses. I am grateful to the following groups that allowed me to hone the narrative: the SPACE TOURS group at the Institut Denis Poisson (France), notably Kilian Raschel and Cedric Lecouvey; the graduate students and faculty at the University of Linz (Austria), especially Veronika Pillwein, Manuel Kauers, Christophe Koutschan, Ellen Wong; the *petite école de combinatoire* of LaBRI, University of Bordeaux (France), especially the questioning of Philippe Duchon, Mireille Bousquet-Mélou, Jean-François Marckert, and Yvan leBorgne. LaBRI most generously hosted me and provided important resources during a substantial portion of the writing.

Thank you to my dear students and other co-authors that have worked on related projects and read portions of this book. Stephen Melczer above all, but also Jason Bell, Julien Courtiel, Andrew Rechnitzer, Mercedes Rosas, Samuel Simon, Sheila Sundaram, Stefan Trandafir and Alexandria Vassallo, in addition to those already mentioned. Discussions with Alin Bostan, Frédéric Chyzak, Igor Pak, Bruno Salvy, Lucia diVizio and Michael Singer were essential to clarifying a number of concepts, particularly around transcendency and generating function classification.

My research and travels have been supported by CNRS, PIMS Europe and an NSERC Discovery Grant.

I dedicate this work to my patient children Madeleine, Eleanor and Felix. Young Felix suggested *Mathematics for Adults* as a title, and indeed I hope this text is accessible to a wide number of adults, not just specialists in the area.

Cedric Chauve was essential to the success of this work from both a personal and professional standpoint. He is patient yet demanding and always supportive. Thank you from deep down.

Finally, I am eternally grateful for inspiration, mentorship and early career support of the mighty *Algorithmix*, Philippe Flajolet.

Sans technique, le talent n'est rien qu'une sale manie.
GEORGES BRASSENS

Marni Mishna
Bordeaux, France

Symbols

Symbol Description

$\mathbb{N}$	The set of natural numbers, $\{0,1,\ldots,\}$
$\mathbb{R}$	The set of real numbers
$\mathbb{C}$	The set of complex numbers
K	A field of characteristic zero
$K[x]$	Polynomials in x with coefficients in K
$K[[x]]$	Formal power series in x with coefficients in K
$K(x)$	Rational functions in x over K
$K[x,x^{-1}]$	Laurent polynomials in x with coefficients in K
$K((x))$	Laurent series in x with coefficients in K
$[x^n]F(x)$	Coefficient of x^n in a Taylor/Laurent series expansion of $F(x)$ around 0.
e_j	Elementary basis vector
$\mathbf{x}$	$(x_1,x_2,\ldots,x_d)$
$\mathbf{x}^\alpha$	$x_1^{\alpha_1}\cdots x_d^{\alpha_d}$
$\mathbf{x}^{-1}$	$(x_1^{-1},\ldots,x_d^{-1})$
$e^{\mathbf{x}}$	$(e^{x_1},\ldots,e^{x_d})$
$T(\mathbf{z})$	Torus of the point $\mathbf{z}=\left\{\mathbf{z}' \mid \left|z_j'\right| = \left|z_j\right|\right\}$
$D(\mathbf{z})$	Polydisk $\left\{\mathbf{z}' \mid \left|z_j'\right| \leq \left|z_j\right|\right\}$
$F(\mathbf{x})$	$\sum_{\mathbf{n}} f(\mathbf{n})\mathbf{x}^{\mathbf{n}}$
$\mathrm{Supp}(F)$	The series support: $\{\mathbf{n} \mid f(\mathbf{n}) \neq 0\}$
$\mathrm{CT}\,F(\mathbf{x})$	The constant term $f(\mathbf{0})$
$\Delta F(\mathbf{x})$	Diagonal series $\sum_n f(n,n,\ldots,n)\,x_d^n$
$\Delta^{\mathbf{r}}F(\mathbf{x})$	Diagonal series along $\mathbf{r}$: $\sum_n f(r_1 n, r_2 n,\ldots,r_n n)\,x_d^n$
$\binom{n}{n_1,\ldots,n_k}$	Multinomial $\frac{n!}{n_1!\ldots n_k!}$
$\mathrm{res}_{z=z_0} F(z)$	The residue of $F(z)$ at z_0
$h(\mathbf{z})$	The height function $-\sum \log|z_i|$
$\mathrm{relog}(\mathbf{z})$	Relog map $(-\log|z_1|,\ldots,-\log|z_d|)$
$\mathcal{H}$	Hessian matrix $\left[\partial_j\partial_k\phi(t)\right]$
$\mathbf{z}^-$	$(z_1,\ldots,z_{d-1})$
$\mathbb{P}$	A convex polytope
$n\mathbb{P}$	The n-th dilate of $\mathbb{P}$
$L_{\mathbb{P}}(n)$	Lattice point enumerator: $\#n\mathbb{P}\cap\mathbb{N}^d$

Welcome to Analytic Combinatorics

The art and science of counting discrete objects is an important aspect of modern applied modelling. Enumeration formulas help us understand the large-scale behaviour of objects; they can confirm or refute whether exceptional structure is expected or not; they allow us to tailor algorithms and data structures for maximal efficiency.

Counting sequences are encoded into the Taylor series known as generating functions. The simple series has proved to be far more than a convenient storage devices for sequences. Structure and complexity results are gleaned from properties of the series viewed as an analytic function. A central development in the last century was the twinning of complex analysis and combinatorics to develop incredibly precise estimates of counting functions. Indeed the invocation of analysis has led to a deluge of methods for counting and random generation in addition to a greater understanding of the hierarchy of combinatorial structure. For example, treating series coefficients as contour integrals avails us of a significant, robust toolbox, now nearly 200 years old. Residue computations facilitate computation considerably.

In this book we consider combinatorial structures and questions about enumeration. We also ask natural taxonomic questions – to what extent is there a natural hierarchy of combinatorial classes, and to what extent can we access this through generating functions? Ideally, we understand underlying context, not just individual combinatorial classes or specific counting sequences. What does it mean for a class to be rational, or algebraic?

The analytic combinatorics framework for asymptotic enumeration proceeds somewhat systematically in three major steps:

A combinatorial description of a class $\mathscr{C}$

Systems of functional equations satisfied by the generating function $C(x)$

Asymptotic counting formulas for the number of objects of size, C_n

We encode the basic property of size in the base variable, and track other combinatorial properties with other variables. Analysis of multivariable complex functions at their singular points gives very precise enumerative information and also distribution information about the parameter values.

> ### Locate the singularities
>
> ### Identify the critical singularities
>
> ### Determine the behaviour at these critical points
>
> ### Sum over contributions

The text is organized in two parts: In the first, we develop combinatorial notions and understanding of properties and families of discrete objects. The second part moves to the analysis of combinatorial generating functions via contour integrals. The main goal is to develop asymptotic formulas for counting sequences.

Symbolic methods First we recall the formalisms of combinatorial structures, and the link between basic combinatorial constructions and the corresponding operations on generating functions.

Multivariate generating functions We consider additional combinatorial parameters, and track this additional information into multivariable generating functions

Derived and transcendental classes We go beyond S-regular and algebraic classes and consider combinatorial classes defined by combinatorial extraction methods. We introduce the diagonal operation on generating functions, which is a convenient method to handle an important family of transcendental classes.

Coefficient asymptotics of univariate generating functions We consider how to extract information about the large-scale (asymptotic) properties of combinatorial objects from the behaviour of their generating functions around their singularities.

Singularity structure of multivariate rational functions Once we understand notions of convergence domains and singularities of multivariable rational functions, we can localize singularities.

Integral theorems for multivariable coefficient asymptotics We write coefficients as generalized Cauchy integrals and show how they can be modified into templates which lead to approximation formulas.

Examples in detail We illustrate the multidimensional methodology on two classic combinatorial problems.

Counting lattice paths restricted to cones

Estimating polytope point enumerators

This text is an introduction to the theory, and as such we focus on pedagogical examples coming from combinatorics. In our simplified setting, the rational functions we encounter are highly structured – they are mainly algebraic combinations of geometric series. The general theory is much richer, and having mastered this case, the reader is encouraged to investigate. Similarly, while it is well within the scope here to discuss analysis and geometry, readers will receive only a glimpse, enough to understand the main features.

More than simply a technique-driven manual, we examine classification schemes both of combinatorial classes and of their generating functions. Insights flow in both directions. Series with positive integer coefficients are special in the world of analytic functions, and we understand some special properties by looking at the combinatorial classes.

Part I

Enumerative Combinatorics

1

A Primer on Combinatorial Calculus

CONTENTS

1.1	Combinatorial Classes		4
1.2	Words and Walks		4
	1.2.1	Words and Languages	5
	1.2.2	Lattice Walks	5
	1.2.3	What Is a Good Combinatorial Formula?	7
	1.2.4	Bijections	7
	1.2.5	Combinatorial Operations	8
1.3	Formal Power Series		11
	1.3.1	Ordinary Generating Functions	12
	1.3.2	Coefficient Extraction Techniques	13
1.4	Basic Building Blocks		14
	1.4.1	Epsilon Class	14
	1.4.2	Atomic Class	14
	1.4.3	Admissible Operators and Generating Functions	15
1.5	Combinatorial Specifications		17
1.6	S-regular Classes and Regular Languages		18
	1.6.1	Finite Automata	19
1.7	Tree Classes		21
	1.7.1	Lagrange Inversion	23
1.8	Algebraic Classes		24
1.9	Discussion		28
1.10	Problems		29

A principle objective of combinatorial enumeration is to count the number of objects of a certain size in a given family. This might be the number of graphs on a fixed number of nodes or the number of arrangements of a set of objects under some symmetry constraints. To approach this in a systematic manner, we build a formalism that starts with a precise notion of a combinatorial class. The intuition is developed using two common families of objects: random walks and formal languages. The study becomes systematic as we define a combinatorial calculus with combinatorial operators that simulate addition, multiplication and a quasi-inverse. We will use this to determine counting formulas in a systematic way using **generating functions.** We will consider some high level techniques that process a wide number of classes systematically and algorithmically.

1.1 Combinatorial Classes

Consider a set of objects $\mathscr{C}$ equipped with a map from $\mathscr{C}$ to $\mathbb{N}$. We say that $\mathscr{C}$ is a **combinatorial class** if for any size n the number of objects of that size is finite. The size of an object $\gamma \in \mathscr{C}$ is denoted $|\gamma|$. If there are multiple classes under consideration, the class is made precise by a subscript, $|\gamma|_{\mathscr{C}}$. Combinatorial classes are thus discrete objects of study, as the size is a nonnegative integer.

Given a class $\mathscr{C}$, the subclass of objects of size n is denoted $\mathscr{C}_n$:

$$\mathscr{C}_n := \{\gamma \in \mathscr{C} : |\gamma| = n\}. \tag{1.1}$$

The cardinality of this set is denoted c_n, and the sequence $c_0, c_1, \cdots = (c_n)_{n=0}^{\infty}$ is the **counting sequence of the combinatorial class** $\mathscr{C}$. The main objective of analytic combinatorics is, given a class $\mathscr{C}$, to understand $\mathscr{C}_n$ for large n, including formulas for c_n. It turns out that in many cases we can proceed systematically. The optimistic mantra of Flajolet and Sedgewick, which we shall embrace here, is

If you can specify it, you can analyze it.

We next illustrate the definitions and the philosophy with some examples.

1.2 Words and Walks

There are two major families of combinatorial classes that we will explore to illustrate a combinatorial framework, and the main analytic techniques: these are formal languages and lattice walks. Do not make the mistake of thinking that these are the only kinds of classes to which the techniques herein apply! From a pedagogical point of view

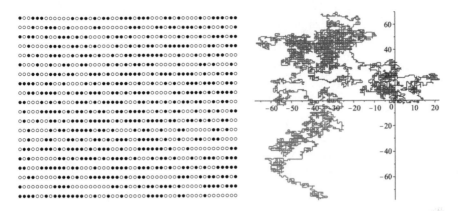

FIGURE 1.1
A large word and a large walk: *(left)* A binary word of size 1000; *(right)*
A simple walk of size 100 000.

the elegance and simplicity of these classes renders them ubiquitous,
and so they are useful to master early.

1.2.1 Words and Languages

Consider a finite set of symbols Σ. If we refer to this as an **alphabet**
we then speak of a **word over** Σ, which is a (possibly empty) sequence
of symbols from Σ. A set of words over Σ is then a **language**. By as-
signing the size of a word to be the length of the sequence, we see
that languages are combinatorial classes. Formal language theory developed in the 21st century as a means to model actual language and
also to formulate instructions for computers.

Example 1.1 (Binary words). Consider $\Sigma = \{\circ, \bullet\}$, and let $\mathscr{B}$ be the set
of all words over Σ. Since the cardinality of Σ is two, $\mathscr{B}$ is a set of
binary words. Figure 1.1 gives a large example. We denote the word
of length 0, the empty word, by ϵ:

$$\mathscr{B} := \{\epsilon, \circ, \bullet, \circ\circ, \circ\bullet, \circ\bullet\circ, \circ\bullet\bullet, \circ\circ\circ, \circ\circ\bullet, \bullet\bullet\circ, \bullet\bullet\bullet, \ldots\}.$$

The words of length 2 form the set $\mathscr{B}_2 = \{\circ\circ, \circ\bullet, \bullet\circ, \bullet\bullet\}$. We write $b_2 =$
4. By elementary counting argument, $b_n = 2^n$. As a side remark: A map
from $\mathscr{B}$ to $\mathbb{N}$ given by the number of $\circ$ in the word does not define a
combinatorial class – there are an infinite number of words with no
$\circ$, for example. We will investigate binary words using this value as a
parameter in the next chapter. ◄

1.2.2 Lattice Walks

Lattice walks are an important family of combinatorial classes as they offer a straightforward encoding of many mathematical objects, yet remain easy to manipulate and to count. They are used to model familiar objects like queues and molecular structures, and also mathematical concepts like multiplicities in representation theory. Meanwhile, their basic recursive structure means that we can apply many mathematical strategies to develop enumerative formulas. The theory of random walks is a vast topic that is able to determine sharp results on the probabilistic large-scale behaviour of many natural phenomena.

The key ingredients to define are: a finite set of vectors $\mathscr{S} \subset \mathbf{Z}^d$, called a **step set**, which defines the allowable movements on the lattice of integers. The size of the walk is the number of steps. If the step set is finite, then the number of walks of length n is finite and is bounded above by $|\mathscr{S}|^n$. Interesting combinatorial classes arise as we restrict steps and regions on the lattice. We can most easily consider walks restricted to various cones (such as half-planes, quadrants or hyperplanes), strips, each imposing different relations between the vectors. Figure 1.1 contains a large example of a **simple walk**, which is unconstrained in the plane. Simple walks have as step set the elementary-basis vectors and their negatives. We denote by e_j the vector, which has a 1 in the jth position and 0s elsewhere.

Example 1.2 (Motzkin paths). The set of **Motzkin paths** are defined on the upper half-plane, $\mathbf{N}^2$ with step set $\{(1,1),(1,-1),(1,0)\}$ (more compactly denoted $\{\nearrow, \searrow, \rightarrow\}$). A walk of length n in this class starts at $(0,0)$, and ends at $(n,0)$, and *never goes below the line $y = 0$*, that is, they are restricted to the upper half-plane. We can exhaustively generate the objects of length 3:

$$\mathscr{M}_3 = \left\{ \nearrow\searrow\rightarrow, \quad \nearrow\overset{\rightarrow}{}\searrow, \quad \rightarrow\nearrow\searrow, \quad \rightarrow\rightarrow\rightarrow \right\}, \qquad (1.2)$$

and deduce that $m_3 = 4$. ◄

Example 1.3 (Simple quadrant walks). A **two-dimensional simple walk** is a sequence of unit steps $\leftarrow, \rightarrow, \uparrow, \downarrow$ corresponding to the elementary-basis vectors $e_1, -e_1, e_2, -e_2$. A walk is **restricted to the first quadrant** if it remains in $\mathbf{N}^2$. For such a walk at any point in the walk the number of $\uparrow$ is superior or equal to the number of $\downarrow$, and the number of $\rightarrow$ is superior or equal to the number of $\leftarrow$.

The subset of simple walks that start and end at the origin is the class of **excursions**, $\mathscr{E}$. For example, $\mathscr{E}_2 = \{\uparrow\downarrow, \rightarrow\leftarrow\}$. By a parity argument, $e_{2n+1} = 0$, for all n.

More generally, the term excursions might refer to any walk with a prescribed start and end point. In the absence of any other precision, we assume the origin for these end points. ◄

1.2.3 What Is a Good Combinatorial Formula?

Given $\mathscr{C}$, our goal is to understand c_n, but what does this mean, anyhow? For example,

$$c_n = \sum_{\gamma \in \mathscr{C}_n} 1$$

is a formula for c_n, but it isn't particularly useful! Sometimes we can find an exact formula, and sometimes it is too technical to be helpful. We might quantify the difficulty of the formula by asking what is the computational complexity of computing c_n upon the input n? The focus in this text is large-scale behaviour of the formula as n tends to infinity. This facilitates comparison of combinatorial structures and makes clear if the number of objects grows linearly, polynomially, exponentially, ... Part II is dedicated to determining asymptotic formulas.

It is useful to define some basic terminology for this purpose. We write that the counting sequence $(f_n)_{n=0}^{\infty}$ is **asymptotically equivalent** to the function $\Phi(n)$ written $f_n \sim \Phi(n)$, if

$$\lim_{n\to\infty} \frac{f_n}{\Phi(n)} = 1. \tag{1.3}$$

In order to articulate bounds on the growth of a counting sequence we use what is called big-Oh notation. We write

$$f(x) = O(\Phi(x))$$

if there is some $M \neq 0$ and some real valued $x_0 > 0$ so that $|f(x)| \leq M\Phi(x)$ when $|x| > x_0$. We use it frequently to denote the higher terms of a series expansion: $1 + x + x^2 + O(x^3)$.

1.2.4 Bijections

Two combinatorial classes, $\mathscr{A}$ and $\mathscr{B}$, are said to be in **combinatorial bijection** if they have the same counting sequence, i.e. $a_n = b_n$, for all n. We denote the existence of a bijection between $\mathscr{A}$ and $\mathscr{B}$ by $\mathscr{A} \equiv \mathscr{B}$. Ideally, there is a meaningful size-preserving map between the objects in the classes when this is the case. We regularly employ bijections to describe objects. It is a great pleasure in combinatorics to find a satisfying bijection between two classes that have no superficial resemblance or common recursive structure, particularly when the bijection imparts new information to each class. Also, it permits us some simplifying shorthands when we describe some classes.

To prove that two classes are in bijection there are a few options. By definition, it is sufficient to show that the two classes possess the same counting sequence, but this is rarely enlightening on its own, indeed it is generally just the starting point. A bijection is satisfying when it is given by a simple size-preserving map between the two underlying sets such that the defining features are related explicitly. Care must be taken to show that such a map is well-defined, surjective and injective.

An established tool for discovering bijections is the Online Encyclopedia of Integer Sequences (OEIS). This is a phenomenal database of sequences where the entries are refereed, and there are many references to follow. The OEIS is located at: http://www.oeis.org. The entry for binary words is A000079, and the one for the Motzkin sequence is A001006.

1.2.5 Combinatorial Operations

Common set operations build new combinatorial classes from old. Any operation must come equipped with a description of how the size is computed in the resulting set and a verification that the number of objects of any fixed given size is finite. We focus on operations with a particular property: We can determine the counting sequence of the resulting class knowing only the counting sequences of the operands. These are called **admissible operations** and permit us to translate combinatorial equations directly into enumeration equations in a very general way.

The usual set union is not admissible, because the cardinality of the union of two sets depends entirely on their contents. Consider that $|\mathscr{A}_n \cup \mathscr{B}_n|$ depends on the size of their intersection, $|\mathscr{A}_n \cap \mathscr{B}_n|$. However, we can still define a useful union operator.

In general, suppose there is a set operation Φ that takes as inputs a finite number ℓ of combinatorial classes $\mathscr{A}^{(1)}, \ldots, \mathscr{A}^{(\ell)}$, and determines a new set:

$$\mathscr{C} = \Phi(\mathscr{A}^{(1)}, \ldots, \mathscr{A}^{(\ell)}).$$

Then Φ is admissible if c_n depends only on the counting sequences of the $\mathscr{A}^{(j)}$ and not the contents of the classes.

Combinatorial Sum

The **combinatorial sum** of two classes, $\mathscr{A}$ and $\mathscr{B}$, is the set formed by their disjoint union, such that size is inherited. To avoid confusion with the usual union, we denote it with a +: $\mathscr{A} + \mathscr{B}$. The size of the object in the sum is given by its size in its originating class, and this is well-defined:

$$|\gamma|_{\mathscr{C}} := \begin{cases} |\gamma|_{\mathscr{A}} & \text{if } \gamma \in \mathscr{A} \\ |\gamma|_{\mathscr{B}} & \text{if } \gamma \in \mathscr{B}. \end{cases} \tag{1.4}$$

The combinatorial sum of two classes that have an intersection leads to duplications of elements, but we retain the information of its source. The sum of a class with itself is in effect a complete duplication of the class.

A relation on the counting sequence follows

$$\mathscr{C} = \mathscr{A} + \mathscr{B} \implies c_n = a_n + b_n.$$

Combinatorial Product

We denote by the symbol $\times$ the **cartesian product** of two classes:

$$\mathscr{A} \times \mathscr{B} := \{(\alpha, \beta) : \alpha \in \mathscr{A}, \beta \in \mathscr{B}\}. \tag{1.5}$$

In many examples the product is equivalent to a concatenation of subclasses – an object of type $\mathscr{A}$ followed by an object of type $\mathscr{B}$. Often we do not write the $\times$ operator, we simply list the classes in sequence: $\mathscr{C} = \mathscr{A}\mathscr{B}$. This is consistent with the usual way in which formal languages are presented.

The size of an element (α, β) is given by the sum of the sizes of the components:

$$|(\alpha, \beta)| := |\alpha| + |\beta|.$$

Thus, the total number of objects of size n is determined from the counting sequence of its composite objects, independent of the content of the classes:

$$\mathscr{C} = \mathscr{A} \times \mathscr{B} \implies c_n = \sum_{k=0}^{n} a_k b_{n-k}.$$

Consequently, the product is an admissible operator.

Powers

Using repeated application of the product operator, we can describe a power operator for whole number powers:

$$\mathscr{A}^\ell := \{(\alpha_1, \ldots, \alpha_\ell) : \alpha_j \in \mathscr{A}\} \equiv \underbrace{\mathscr{A} \times \cdots \times \mathscr{A}}_{\ell \text{ times}}.$$

Since the product is admissible, so is the power operator. We can describe the counting sequence explicitly using compositions of n:

$$\mathscr{C} = \mathscr{A}^\ell \implies c_n = \sum_{k_1 + \cdots + k_\ell = n} a_{k_1} \ldots a_{k_\ell}.$$

We take the convention that $\mathscr{A}^0$ is the class of a single element of size 0. This is called an epsilon class, and we will say more about this in a moment.

Sequence

To express the idea of powers of arbitrary length, we define a **sequence operator**, also known as Kleene star:

$$\mathscr{A}^* = \text{Seq}(\mathscr{A}) := \bigcup_{k \geq 0} \mathscr{A}^k.$$

Note, the definition is a sum over an infinite number of powers, and thus to ensure that we construct a combinatorial class, we can only apply it when $\mathscr{A}_0 = \emptyset$, else the resulting set contains an infinite number of elements of size 0. When we consider an element from $\mathscr{A}^\ell$, we say that the length of the the the sequence is ℓ.

Again, the operation is clearly admissible. It is less practical to describe in terms of the counting sequence. We move next to a more convenient encoding to describe the action on the counting series.

1.3 Formal Power Series

To harness the power of analysis to study combinatorial classes, we encode their counting sequences in the coefficients of formal power series. These are objects from the ring $\mathbb{Z}[[x]]$. This permits a very powerful shorthand for counting sequences, as we identify functions and their Taylor series at the origin.

A **formal power series** $\sum_{n=0}^{\infty} f_n x^n$ is a limit of partial finite sums. We can consider the ring of formal power series, where the sum is pointwise, and the product is given by the natural extension of polynomial multiplication:

$$\sum f_n x^n \sum g_n x^n = \sum_n \left(\sum_k f_k g_{n-k} \right) x^n.$$

The element $1 - x$ is has a multiplicative inverse in the ring. It is the called the **geometric series**:

$$\frac{1}{1-x} = \sum_{n=0}^{\infty} x^n. \tag{1.6}$$

We are interested in the **Taylor series of a function $F(x)$ at a point x_0** defined (when it exists) as the series:

$$\sum_{n \geq 0} \frac{1}{k!} \frac{d^k}{dx^k} F(x_0)(x - x_0)^k.$$

The geometric series is the Taylor series of $\frac{1}{1-x}$.

This ring is of interest to us as there are numerous formal manipulations which correspond to analytic operations. Let $F(x) = \sum_{n=1} f_n x^n$, then

$$F'(x) := \sum_{n=0} f_{n+1} x^n \quad \text{and} \quad \int F(x)\, dx := \sum_{n=1} f_{n-1} \frac{x^n}{n}.$$

The composition of two series $F(G(x))$ is well defined as $\sum_n f_n (G(x)^n)$ provided that either $F(x)$ is a polynomial or $G(0) = 0$.

1.3.1   Ordinary Generating Functions

Given a combinatorial class $\mathscr{C}$, the **ordinary generating function** or **OGF** of the class is the formal power series:

$$C(x) = \sum_{\gamma \in \mathscr{C}} x^{|\gamma|}, \quad \text{equivalently} \quad C(x) = \sum_{n \geq 0} c_n x^n. \tag{1.7}$$

Example 1.4. For the class of binary words, we know a priori that $b_n = 2^n$, and hence its generating function is a geometric series:

$$B(x) = \sum_{n=0}^{\infty} b_n x^n = \sum_{n=0}^{\infty} 2^n x^n = \frac{1}{1 - 2x}. \tag{1.8}$$

At this point it is not clear just yet which is the better way to store the information.   ◀

Given *any* sequence of numbers, we can speak of the generating function of the sequence, but this might not gain much. In the course of proofs, we frequently make use of both the non-negativity and the integrality of the series coefficients. For example, you will note that the fact that a generating function increases as x tends to infinity is frequently used in the proofs.

Two other types of generating functions are common in combinatorics, and although not the focus here, you will encounter them. The **exponential generating function** (EGF) of a sequence is the series $\sum_{n \geq 0} c_n \frac{x^n}{n!}$. This is the better choice for combinatorial classes which bear some sort of labelling, or permutation structure, as the additional factorial assures series convergence. Indeed, it is not clear what can be done with $\sum n! x^n$, after all. In this text, we focus on unlabelled objects and hence restrict our attention to ordinary generating functions.

A **probability generating function** (PGF), is a the generating function for a discrete probability distribution. More precisely, they are of the form $\sum p_n x^n$, such that $p_n \geq 0$ and $\sum_n p_n = 1$.

We use the following slightly strange but common notation to describe coefficient extraction:

$$[x^n]\left(\sum f_k x^k\right) := f_n.$$

For example, $[x^3]2x+4x^2+x^3 = 1$ and $[x^4]2x+4x^2+x^3 = 0$. The notation $[x^n]F(x)$, for some function $F(x)$, implies the coefficient of x^n in the Taylor series expansion of $F(x)$ about the origin. (Or, any other relevant series expansion.) For example, $[x^n]\frac{1}{1-2x} = 2^n$ and $[x^k](1-x)^n = \binom{n}{k}$ for integer n.

1.3.2 Coefficient Extraction Techniques

To extract coefficients, we have a few basic identities that we can use:

$$[x^n](F(x) + G(x)) = ([x^n]F(x)) + ([x^n]G(x))$$

$$[x^n]F(\rho x) = \rho^n[x^n]F(x).$$

Newton's generalized binomial theorem is a valuable tool for exact coefficient extraction. The **extended binomials** generalize binomials. For $k \in \mathbb{N}$ and $r \in \mathbb{R}$ define:

$$\binom{r}{k} := \frac{r(r-1)(r-2)\ldots(r-k+1)}{k!}.$$

Then

$$[x^k](1+x)^r = \binom{r}{k}, \tag{1.9}$$

When the power is negative, it is convenient to reorganize this expression. Let $r \in \mathbb{R}^+$. Then, for any real r and nonnegative integer k,

$$\binom{-r}{k} = \frac{(-r)(-r-1)\ldots(-r-k+1)}{k!} = (-1)^k\binom{r+k-1}{k}. \tag{1.10}$$

We can apply this to deduce

$$[x^n]\frac{1}{(1-mx)^r} = m^n\binom{r+n-1}{n} \tag{1.11}$$

for real numbers m and r. We can build an exact expression for any rational function series coefficient. We first rewrite this rational function $R(x)$ into the form $R(x) = \sum_{i,k} \frac{\alpha_{i,k}(x)}{(1-m_i x)^k}$, such that each $\alpha_{i,k}(x)$ is a polynomial of degree less than k, then we can expand each term and apply Eq. (1.11).

Example 1.5 (Simple rational example). In order to determine the coefficient of x^n in the Taylor series expansion of

$$R(x) = \frac{486x^6 - 810x^5 + 540x^4 - 180x^3 + 14x^2 + 6x - 1}{(-1+3x)^5(-1+4x)^2},$$

first, compute a partial fraction decomposition:

$$R(x) = \frac{1}{(1-3x)^5} + \frac{2x}{(1-4x)^2}.$$

We apply Eq. (1.11) to each term in the sum

$$
\begin{aligned}
[x^n]R(x) &= [x^n]\frac{1}{(1-3x)^5} + [x^n]\frac{2x}{(1-4x)^2} \\
&= 3^n\binom{4+n}{n} + 2\,4^{n-1}\binom{n}{n-1} \\
&= 3^n(n+4)(n+3)(n+2)(n+1) + 2n4^{n-1}.
\end{aligned}
$$

In this case, we can conclude the asymptotic estimate $[x^n]R(x) \sim \frac{4^n n}{2}$.

◂

1.4 Basic Building Blocks

We can describe combinatorial classes using some basic building blocks. To this end, we introduce the **epsilon** classes and **atomic** classes, which contain objects of size 0 and 1, respectively.

1.4.1 Epsilon Class

An epsilon class contains a single element of size 0. Most formally, it is denoted $\mathcal{E} = \{\epsilon\}$, with $|\epsilon| = 0$, but most often written with the shorthand ϵ to denote the set that contains it. This is akin to the ϵ word in formal language theory. Since it is a class with a single object of size 0, we deduce that $E(x) = 1$. We have seen the power $\mathcal{A}^0$ is an epsilon class. Epsilon classes can be used as base cases for recursive constructions. They can also be used to track internal structure of an element, for clarity or counting purposes.

1.4.2 Atomic Class

An atomic class is a set that contains a single element of size 1. Again, formally, we write $\mathcal{Z} = \{\circ\}$ where $|\circ| = 1$, but it is often clearer to represent the set containing the element by the element itself. The generating function of a single object of size one is $Z(x) = x$. The atomic elements are the building blocks that give the size. This could be a letter in the alphabet for a word class, or a single step of step set, the nodes of a graph, to name just a few simple examples.

TABLE 1.1

Elementary Combinatorial Operations and Their Generating Function Implications

$\mathscr{C}$	$C(x)$
Atomic Class	x
Epsilon Class	1
$\mathscr{A} + \mathscr{B}$	$A(x) + B(x)$
$\mathscr{A} \times \mathscr{B}$	$A(x)B(x)$
$\mathscr{A}^k$	$A(x)^k$
$\mathscr{A}^*$	$\frac{1}{1-A(x)}$ (provided $a_0 = 0$)

1.4.3 Admissible Operators and Generating Functions

The main advantage of describing admissible operators is that we can build an automatic mechanism to translate combinatorial equations into generating function equations. Each of the admissible operations we have seen so far has a natural action on the generating function.

Theorem 1.1. *Let $\mathscr{A}, \mathscr{B}$ and $\mathscr{C}$ be combinatorial classes. The functional relations summarized in Table 1.1 are true.*

Proof. For combinatorial sum, product and power, the generating function relations follow from the counting sequence formulas. Here are sum and product:

$$\sum_{\gamma \in \mathscr{A} + \mathscr{B}} x^{|\gamma|} = \sum_{\gamma \in \mathscr{A}} x^{|\gamma|} + \sum_{\gamma \in \mathscr{B}} x^{|\gamma|} = A(x) + B(x)$$

$$\sum_{(\alpha, \beta) \in \mathscr{A} \cdot \mathscr{B}} x^{|\alpha| + |\beta|} = \sum_{\alpha \in \mathscr{A}} \sum_{\beta \in \mathscr{B}} x^{|\alpha|} x^{|\beta|} = A(x)B(x).$$

For the sequence construction, the result is a consequence of relation $(1 - x) \sum x^n = 1$, and the fact that composition of formal power series is well defined here as $a_0 = 0$. $\square$

Example 1.6 (Binary words several ways). A binary word is a sequence of atoms, of which there are two types: $\circ$ and $\bullet$. We can describe the class as $\mathscr{B} = \{\circ, \bullet\}^*$. We translate the combinatorial objects into generating functions as follows:

$$\{\circ, \bullet\} \mapsto 2x \qquad \{\circ, \bullet\}^* \mapsto \frac{1}{1 - 2x}$$

We apply the sequence rule of Theorem 1.1 to determine $B(x) = \frac{1}{1-2x}$. For an alternative derivation, we could say that a binary word is either empty, or it is an atom in the alphabet, followed by another binary word:

$$\mathcal{B} \equiv \{\epsilon\} + \{\circ, \bullet\} \times \mathcal{B}.$$

We can apply Theorem 1.1 to deduce a functional equation for the generating function:

$$
\begin{array}{ccccccc}
\mathcal{B} & \equiv & \{\epsilon\} & + & \{\circ, \bullet\} & \times & \mathcal{B} \\
\downarrow & & \downarrow & & \downarrow & & \downarrow \\
B(x) & = & 1 & + & 2x & \cdot & B(x).
\end{array}
$$

This gives an alternative decomposition for the class, but of course the same generating function results:

$$B(x) = 1 + 2x\,B(x) \implies B(x) = \frac{1}{1 - 2x}.$$

Note: The translations hold even in recursive statements. We can also uniquely decompose a binary word into subwords starting with a $\bullet$. We have simplified some notation (using brackets to illustrate order instead of several sets of nested braces):

$$\mathcal{B} \equiv \circ^* \times (\bullet \times (\circ)^*)^* \implies B(x) = \frac{1}{1-x}\frac{1}{1-\frac{x}{1-x}} = \frac{1}{1-2x}.$$

While this form initially appears unnecessarily complicated, it is particularly amenable to studying forbidden patterns in binary strings. ◄

Example 1.7 (Motzkin paths)**.** Consider a lattice walk defined by the step set $\{\rightarrow, \nearrow, \searrow\}$. We can uniquely decompose a Motzkin path by its first return to height 0. That is, a Motzkin path is either empty, or it is a non-empty path that returns to the axis, followed by another (possibly empty) Motzkin path:

$$\mathcal{M} \equiv \epsilon + \rightarrow \mathcal{M} + \nearrow \mathcal{M} \searrow \mathcal{M}. \qquad (1.12)$$

We deduce a recursive generating function equation:

$$
\begin{array}{ccccccc}
\mathcal{M} & \equiv & \epsilon & + & \rightarrow \mathcal{M} & + & \nearrow \mathcal{M} \searrow \mathcal{M} \\
\downarrow & & \downarrow & & \downarrow & & \downarrow \\
M(x) & = & 1 & + & xM(x) & + & xM(x)xM(x).
\end{array}
$$

The functional equation $M(x) = xM(x) + x^2 M(x)^2 + 1$ is easily solved using the quadratic formula. There are two solutions. We choose the one with the correct Taylor series expansion, i.e. the one that is 1 at $x = 0$:

$$M(x) = \frac{1 - x - \sqrt{-3x^2 - 2x + 1}}{2x^2} = 1 + x + 2x^2 + \dots \qquad (1.13)$$

From this expression to find m_n, there are a few options. To go from generating function to coefficient, we can apply the generalized binomial theorem (after some initial manipulation). Alternatively, we can use the analytic methods of Chapter 4 to compute an asymptotic estimate. ◄

1.5 Combinatorial Specifications

Generically, we can specify a combinatorial class by a list of combinatorial equations:

$$\mathscr{C}_1 = \Phi_1(\mathscr{Z}, \mathscr{C}_1, \dots, \mathscr{C}_r)$$

$$\vdots \qquad (1.14)$$

$$\mathscr{C}_r = \Phi_r(\mathscr{Z}, \mathscr{C}_1, \dots, \mathscr{C}_r),$$

where each Φ_i is an admissible construction.

The dependency graph of a specification is defined with vertex set $\{\mathscr{C}_j\}$ with an edge from $\mathscr{C}_j$ to $\mathscr{C}_k$ if $\mathscr{C}_k$ is used in Φ_j. That is, to expand $\mathscr{C}_j$, we require a $\mathscr{C}_k$ object.

A specification is **iterative** if the dependency graph is a acyclic. It is **recursive** otherwise. A class could be specified by both – binary words are defined by $\mathscr{B} = \{\circ, \bullet\}^*$ and $\mathscr{B} = \{\circ, \bullet\}\mathscr{B} + \epsilon$; the former is iterative, and the latter is recursive.

It is important to make sure that the specifications are well-defined. For example, the system $\mathscr{A} = \mathscr{B} + \mathscr{A}, \mathscr{B} = \{\circ\}$ does not lead to a meaningful definition of $\mathscr{A}$. Testing that a class is well-defined can be accomplished using some analytical tests, we will return to this idea in the Discussion.

There are some families of classes that have commonalities in their specifications, and this leads to important characterizations of their generating functions.

1.6   S-regular Classes and Regular Languages

An iterative specification that only involves atoms, combinatorial sums, cartesian products and sequence constructions is said to be a **regular specification**. A regular language can be specified by a regular specification.

A combinatorial class is said to be specification-regular (or **S-regular**) if it is combinatorially isomorphic to a class of objects with a regular-specification. S-regular classes are fundamental in elementary theoretical computer science as they capture an important class of computer languages.

Example 1.8. The class of binary words avoiding a ∘∘ subpattern can be expressed by the following specification:

$$\mathscr{B} \equiv \circ^* \times (\bullet \times (\epsilon + \circ))^*.$$

As $\mathscr{B}$ is defined only in terms of atomic classes, we see that it is iterative. As there are only sequences, sums and products, we can conclude that the specification is S-regular.   ◄

The first thing we can say is a direct consequence of Theorem 1.1:

Theorem 1.2. *Any S-regular language has a rational generating function.*

You can prove this without too much difficulty by induction: Atoms and epsilon classes have polynomial, hence rational generating functions. The finite sum, finite product and quasi-inverse of rational functions are all rational functions, hence an S-regular class has a rational generating function.

A harder question is the inverse: For which rational functions $F(x)$ can you derive a regular specification for a class whose generating function is precisely $F(x)$? Indeed you can (with some work! [Kou08]) find a series in $\mathbb{N}[[x]]$, that is the Taylor series of a rational function that is provably NOT a generating function for an S-regular class.

If you know a class has a non-rational generating function, it cannot be S-regular. For example, the generating function of Motkzin

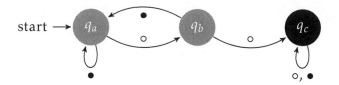

FIGURE 1.2
An automata that reads in strings of $\circ$ and $\bullet$. The accepting states are q_a and q_b, and the rejecting state is q_c.

paths is not rational, thus, no matter how hard we try, we will never find a regular specification for Motzkin paths. It is somehow reassuring that we can conclude this so easily.

1.6.1 Finite Automata

Finite automata offer a graphical description of a regular language. They are arranged into states which process input strings. The processing always begins in a designated state, and the input string is processed and accepted if it ends in an accepting state. Formally, a **finite state machine** is a five-tuple (σ, Q, S, χ, T), where Σ is the input language, the set of all states is Q, $S \in Q$ is a start state, and $T \subset Q$ is a non-empty set of accepting terminating states. Any state in $Q \setminus \{T\}$ is therefore a rejecting state. The function $\chi : (\Sigma, Q) \to Q$ is a transition function and takes an input and a state, and it then indicates a new state.

An automaton is typically represented by a graph with vertex set Q, and (directed) edge set

$$\{(Q_1, Q_2) : \exists(\sigma, Q_1) \mapsto Q_2\}.$$

Example 1.9. Figure 1.2 illustrates a finite automaton. Here, $\Sigma = \{\circ, \bullet\}$, $Q = \{q_a, q_b, q_c\}$, $S = q_a$, and $T = \{q_a, q_b\}$ (the grey states). The transition function is illustrated directly on the edge labellings. For example, $\chi(\circ, q_a) = q_b$. The language accepted by this automaton is the set of binary words with no $\circ\circ$ subword. ◄

Theorem 1.3. *A language can be characterized by regular grammar if and only if there exists a finite automaton which accepts the language.*

The proof is constructive and appears in most introductory texts on formal language theory. It is useful to know this, as it helps us understand that S-regular classes have a finite memory. This can be made more precise and turned into a criterion to prove that a class is not S-regular. A classic strategy in language theory to prove that a language is not regular is to apply a result known as the pumping lemma. We state the result and then illustrate how to use it.

Theorem 1.4 (Pumping lemma for regular languages). *For any regular language $\mathcal{L} \subset \Sigma^*$, there exists an integer n such that for any $w \in \mathcal{L}$ with $|w| \geq n$, there exist words $s, t, u \in \Sigma^*$ so that $w = stu$ and*

1. $|st| \leq n$

2. $|t| \geq 1$

3. *for all $k \geq 0$, $st^k u \in \mathcal{L}$.*

The name comes from the concept that the string t can be "pumped" and the word remains in the language. This characterization is typically used to show that a languages is not regular, by giving $w = stu$ and illustrating for all factorizations there is a k so that the word $st^k u$ is not in the language.

Example 1.10. We illustrate how to apply this result with an example. Consider the language $\mathcal{L} = \{0^n 1^n : n \in \mathbb{N}\}$. The generating function is $L(x) = \sum_n x^{2n} = \frac{1}{1-x^2}$ rational, but we can illustrate that it does not satisfy the pumping lemma, and hence the language is not regular.

Fix any $n \geq 1$. Let stu by any decomposition of $w = 0^n 1^n$. There are a few possibilities. If $t = 0^m$, then $m \geq 1$ since t is not empty and $st^0 u$ will have fewer zeroes than ones contradicting (3) for $k = 0$. A similar argument holds if $t = 1^m$. If t is of the form $t = 0^{m_1} 1^{m_2}$ with m_1, m_2 greater than zero. It follows that $st^2 u = 0^n 1^{m_2} 0^{m_1} 1^n \neq \mathcal{L}$. We conclude that as we can contradict the pumping lemma, $\mathcal{L}$ is not a regular language.   ◄

Languages that are S-regular have several closure properties. The union of a finite number of regular languages is a regular language. The concatenation of a finite number of regular languages is a regular language. The finite intersection of regular languages is regular. Many closure properties are easiest proved by building automata, and others are easier proved from the grammar characterization.

1.7 Tree Classes

Rooted trees can represent structured information in a very natural way. As combinatorial classes they are easily specified, and analyzed. The root of the tree is the starting point for a recursive definition: A tree is a root node connected to a collection of subtrees. A **plane tree** has the property that the subtrees of any node are in a fixed embedding in the plane – they occur in a specified order, which we can capture with products and sequences.

We start with a simple example and see how it can generalize.

Example 1.11 (Binary trees). A **binary tree** is a plane tree such that each node either has zero or two children. The size of a binary tree is the number of nodes. If we remove the root of a binary tree, either we are left with nothing (because it was the tree comprised of a single node), or we are left with two subtrees: a left one and a right one. This gives us a decomposition of a binary tree:

This structure is captured in the following specification:

$$\mathcal{B} = \mathcal{Z} + \mathcal{Z} \times \mathcal{B} \times \mathcal{B}.$$

We translate the classes an operators into a generating function equation:

$$
\begin{array}{ccccccccc}
\mathcal{B} & = & \mathcal{Z} & + & \mathcal{Z} & \times & \mathcal{B} & \times & \mathcal{B} \\
\downarrow & & \downarrow & & \downarrow & & \downarrow & & \downarrow \\
B(x) & = & x & + & x & \cdot & B(x) & \cdot & B(x)
\end{array}
$$

The equation $B(x) = x + xB(x)^2$ is solved using the quadratic formula:

$$B(z) = \frac{1 - \sqrt{1 - 4x^2}}{2x}.$$

Important remark! There is a second solution to this equation, but it does not yield a power series solution. Combinatorial algebraic equations will always have one power series solution. In this case, we can show that the other solution diverges at 0, whereas the power series solution has $B(0) = 0$.

Here x counts the total number of vertices. However, it is more typical (especially in computer science) to count the number of **internal** vertices. We can modify the specification so that leaves are tagged with an epsilon class instead of an atom:

$$\mathscr{B} = \epsilon + \circ \times \mathscr{B}^2 \implies B(x) = 1 + B(x)^2 = \frac{1 - \sqrt{1 - 4x}}{2}$$

◄

A node in a **rooted plane tree** can have any number of children. By deleting the root of a plane tree we obtain an ordered *sequence* of (smaller) rooted plane trees. The subtrees of a given node form a sequence:

$$\mathscr{G} = \circ \times \mathrm{SEQ}(\mathscr{G}).$$

The translation to generating functions gives:

$$G(x) = \frac{x}{1 - G(x)} \implies G(x) = x + G(x)^2 = \frac{1 - \sqrt{1 - 4x}}{2}.$$

Remarkably, it is the same generating function! We can conclude immediately that binary trees where size is the number of internal vertices are in a combinatorial bijection with general plane trees, counted by number of nodes.

Let us generalize the possible number of children allowed. Let Ω be a set of $\mathbb{N}$ that contains zero. We can define the class of Ω-**restricted trees**, denoted $\mathscr{T}^\Omega$ to be the set of rooted plane trees whose node outdegrees lie only in Ω.

The class of $\mathscr{T}^\Omega$ trees satisfies the following combinatorial equation:

$$\mathscr{T}^\Omega = \circ \cdot \left(\sum_{k \in \Omega} \mathscr{T}^k \right).$$

Any class of trees that satisfies such an equation is called a **simple variety of trees**.

The following theorem about the OGF of $\mathscr{T}^\Omega$ follows immediately from the specification.

TABLE 1.2
Ω-restricted Trees

Ω	$\phi(u)$	Tree type
$\{0, 2\}$	$1 + u^2$	binary trees
$\{0, 1, 2\}$	$1 + u + u^2$	unary binary trees
$\{0, 3\}$	$1 + u^3$	ternary trees
$\mathbb{N}$	$\frac{1}{1-u}$	general plane trees

Lemma 1.5. *Let Ω be a set of $\mathbb{N}$ that contains zero and define*

$$\phi(u) := \sum_{\omega \in \Omega} u^{\omega}.$$

The OGF $T^{\Omega}(x)$, of the class of Ω-restricted plane trees $\mathscr{T}^{\Omega}$, satisfies the following functional equation

$$T^{\Omega}(x) = x\phi(T^{\Omega}(x)).$$

Proof. We decompose any $\tau \in \mathscr{T}^{\Omega}$ by deleting the root vertex. When we do so we must obtain a sequence of trees – in particular, an $\text{SEQ}_{\Omega}(\mathscr{T}^{\Omega})$. The generating function of this sequences is given precisely by ϕ. The result follows from Theorem 1.1. $\qquad\square$

Table 1.2 illustrates how to describe some common trees as Ω-trees.

1.7.1 Lagrange Inversion

Clearly, not every functional equation can be solved using the quadratic formula. There are some other strategies, however. The tree functional relation implies

$$z = \frac{T}{\phi(T)}.$$

If we view T as a function: evaluated at a number z we get $T(z)$. Remarkably, this expression gives a functional inverse: if we evaluate $T/\phi(T)$ at z, we get z. This form can be exploited in order to get an exact expression for T_n – this uses something called the Lagrange inversion formula.

Theorem 1.6 (Lagrange inversion). *Given a functional equation $z = T/\phi(T)$ then*

$$[z^n]T(z) = \frac{1}{n}[\omega^{n-1}]\phi(\omega)^n$$

$$[z^n]T(z)^k = \frac{k}{n}[\omega^{n-k}]\phi(\omega)^n.$$

Note that this immediately gives

$$T_n^\Omega = \frac{1}{n}[\omega^{n-1}]\phi(\omega)^n.$$

Example 1.12 (Plane trees). To determine the exact number of plane trees of size n, we apply the Lagrange inversion formula:

$$T(z) = \frac{z}{1 - T(z)} \qquad\qquad \phi(u) = \frac{1}{1 - u}$$

$$\phi(u)^n = \frac{1}{(1 - u)^n} = \sum_{k=0}^{\infty}\binom{n + k - 1}{k}u^k$$

$$T_n = \frac{1}{n}[u^{n-1}]\phi(u)^n = \frac{1}{n}\binom{2n - 2}{n - 1}.$$

◀

1.8   Algebraic Classes

We can recast trees in a more general setting to consider those classes which are specified by a system of equations.

A combinatorial class $\mathcal{C}$ is said to be **algebraic** if it can be written as the first component $\mathcal{C} = \mathcal{S}$ of a system of equations

$$\mathcal{S}_1 = \Phi_1(\mathcal{Z}, \mathcal{S}_1, \ldots, \mathcal{S}_r)$$

$$\vdots \qquad\qquad\qquad (1.15)$$

$$\mathcal{S}_r = \Phi_r(\mathcal{Z}, \mathcal{S}_1, \ldots, \mathcal{S}_r),$$

where each Φ_i is a construction that only involves $+, \times$ and the neutral class.

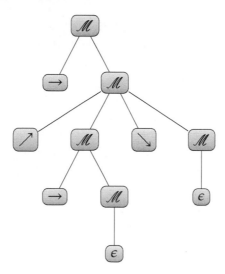

FIGURE 1.3
The derivation tree for the Motzkin Path $_\rightarrow\nearrow^{\rightarrow}\searrow$.

A structure in an algebraic class is identified with the history of the rules expanded to reach it, expressed as a **derivation tree**. Figure 1.3 gives an example. In order to translate to a generating function, it must be unique for a single object. In this case, the specification is said to be unambiguous. Regular classes are algebraic, however, the set of algebraic classes is strictly larger. Many families of trees are algebraic classes, and as their generating functions are not rational, we know they cannot be regular.

Example 1.13 (Trees). The set of binary plane trees is an algebraic class $\mathscr{B} = \circ + \circ \times \mathscr{B} \times \mathscr{B}$, and the set of general planar trees is also algebraic. We can always rewrite sequence constructions using $+$ and $\times$. For example, $\mathscr{G} = \circ \times \mathrm{S}_{EQ}(\mathscr{G})$ can be rewritten

$$\mathscr{G} = \circ \times \mathscr{F} \qquad \mathscr{F} = \epsilon + \mathscr{G} \times \mathscr{F}.$$

◁

Example 1.14 (Motzkin paths). Consider the Motzkin path specification

$$\mathscr{M} = \epsilon + \, \rightarrow \mathscr{M} + \nearrow \mathscr{M} \searrow \mathscr{M}.$$

This is an algebraic specification. The path is recovered by scanning the leaves of the tree from left to right as demonstrated in Figure 1.3.

◁

Algebraic classes are well-studied in theoretical computer science. Languages that are specified by an algebraic grammar are known as **context-free languages**. A language is said to be an **unambiguous** if every element is derived in a unique way. The label **context-free** comes from formal linguistics, in which the objects can be freely generated from the system of constructions without any external constraints. Herein, context-free will refer to languages, with the understanding that an algebraic object can be viewed as a word on the atoms in a unique manner by reading the leaves of the derivation tree from left to right.

The system of constructions leads naturally to a system of algebraic equations for the generating functions:

$$S_1(x) = \Psi_1(x, S_1(x), \ldots, S_r(x))$$

$$\vdots$$

$$S_r(x) = \Psi_1(x, S_1(x), \ldots, S_r(x)).$$

Using elimination methods from polynomial algebra, we can determine a single polynomial $P \in \mathbb{Q}[x, y]$, such that $P(x, S_1(x)) = 0$. Specifically, one can use resultants or Gröbner basis techniques to eliminate the extraneous variables one-by-one leaving only a polynomial equation for $S_1(x)$ in terms of itself and x.

Theorem 1.7. *An algebraic combinatorial class $\mathscr{C}$ has an ordinary generating function that is an algebraic function. That is there is some nontrivial bivariate polynomial $P(x, y)$, such that*

$$P(x, C(x)) = 0.$$

The characterization of algebraic functions is a classic topic in mathematics, and we will see how to use properties of algebraic functions to prove that a combinatorial class cannot be algebraic because its generating function is not algebraic.

A second approach to proving that a class is not algebraic is a direct argument, analogous to what we saw in the case of the regular languages. It is a characterization developed in the theory of formal languages used to prove that a given language is not a context-free language.

Theorem 1.8 (Pumping lemma for context-free languages). *A language $\mathcal{L} \subset \Sigma^*$ is context-free if there exists an integer $n \geq 1$ (the pumping length), such that every string w in $\mathcal{L}$ of length at most n can be written $w = rstuv$ with substrings $r, s, t, u, v \in \Sigma^*$ satisfying*

1. *$|su| \geq 1$,*

2. *$|stu| \leq n$, and*

3. *$rs^k tu^k v \in \mathcal{L}$ for all $k \geq 0$.*

Example 1.15. Again, we illustrate how to use the theorem by presenting an example. We show that $\mathcal{L} = \{0^n 1^n 2^n : n \in \mathbb{N}\}$ is not a context-free language. The generating function is rational: $L(z) = \frac{1}{1-z^3}$, illustrating that the notions of generating function complexity and language complexity are not perfectly aligned.

To apply the theorem, we let n be arbitrary, and give a word, its decomposition, and a choice of k to contradict point (3) in the theorem statement. For a given $n \geq 1$, let $w = 0^n 1^n 2^n$. We consider the possibilities for stu. The hypotheses guarantee that one of s and u are not both empty.

1. If either or both s and u are words made from a single letter, then as $|su| \geq 1$, $s^0 tu^0 = t$ will have fewer occurrences of at least one and at most two of those letters, and hence $rs^k tu^k v \notin \mathcal{L}$ when $k = 0$.

2. If s or u is not composed of a single letter, then when $k > 1$, the word will not be of the correct form.

In either case, for all n, there exists a word w, such that for any decomposition of w into the correct form, there exists a k so that the pumped version of the word is not in the language. Thus, the language is not context-free. ◄

It is natural to also ask the reverse question: If a combinatorial class has an algebraic generating function, are there criteria to decide when it is an algebraic combinatorial class? Chomsky and Schützenberger proved that given a series $F(x) \in \mathbb{Q}[[x]]$ that is algebraic over $\mathbb{Q}$, one can construct a pair of languages $\mathcal{L}_1$ and $\mathcal{L}_2$, such that $F(x) = L_1(x) - L_2(x)$. The construction is quite natural and amounts to a first grouping of terms to ensure positive coefficients, and then a direct translation. As the construction does not ensure that $\mathcal{L}_2 \subseteq \mathcal{L}_1$, the process does not output a single language with $F(x)$ as generating function.

Indeed, it may be non-trivial to find a natural specification of the form Eq. 1.15. See Exercise 1.13 at the back of the chapter.

Clearly the family of algebraic classes is closed under finite combinatorial sum and concatenation. Context-free languages are closed under intersection with regular languages.

In Chapter 5 we will discuss criteria on algebraic series with positive integer coefficients that provably do not come from context-free grammars.

1.9   Discussion

Formal language theory is a vast theme in theoretical computer science. Specifically within it, the combinatorics of words is a rich subject at the interface of noncommutative algebra, combinatorics and computing. The series of Lothaire [Lot83] is an excellent introduction for those that are combinatorially minded.

There are several formalities which consider different types of set operators and their generating function consequences. The translation of algebraic grammars into a system of equations is often referred to as the DSV methodology (Delest, Schützenberger and Viennot) [Del94]. It is central in the system of decomposable structures and was primarily developed in the late 1980s described by Flajolet, Salvy and van Cutsem. In this formalism, there is a distinction between operators in both the labelled and unlabelled universe, and there is a comprehensive development of sets and cycles operators, which are defined using equivalence classes on sequences and turn out to be admissible. One of the main motivations is to build very well-known objects like partitions, permutations and surjections using a single, automatic formalism. The text *Analytic Combinatorics* [FS09] of Flajolet and Sedgewick outlines this system and the numerous deep consequences.

The theory of species [BLL98] generalizes the notion of equivalence classes using category theory. It permits a very wide variety of class operations. The scope for constructions is larger than decomposable structures, albeit in a more abstract term. The formalism is extremely useful for proofs. For example, Joyal's implicit species theorem gives conditions under which a square system of combinatorial equations admits a unique combinatorial solution. This idea is developed and generalized by [PSS12]. The methods of Pivoteau et al. give an effective Newton iteration to generate objects in a class from a specification. Like Newton's iteration, it has a quadratic convergence, meaning here that after n iterations, you can expect to have generated all the objects of size n^2. In this context, it is also possible to describe a combinatorial derivative, which can be useful to interpret some classes of differential equations.

1.10   Problems

Exercise 1.1 (Catalan numbers in combinatorics). Let $\mathscr{D}$ be the combinatorial class of **Dyck paths:** This is the set of (possibly empty) lattice walks that start at $(0,0)$, take steps $\nearrow = (1,1)$ and $\searrow = (1,-1)$. They end on the x-axis, and never go below it. The length of a walk is the number of steps. Let $\mathscr{D}$ be the combinatorial class of Dyck paths.

1. Prove that $\mathscr{D} = \epsilon + \nearrow \mathscr{D} \searrow \mathscr{D}$. Be sure to justify why the decomposition describes each path in a unique manner.

2. Deduce that the generating function is

$$D(x) = \frac{1 - \sqrt{1 - 4x}}{2x} \quad \text{(OEIS A000108)}.$$

3. The coefficients d_n are **Catalan numbers.** Use the extended binomial theorem to show

$$[x^{n-1}] \frac{1 - \sqrt{1 - 4x}}{2x} = \binom{2n-1}{n-1} \frac{1}{n}.$$

4. Prove that Dyck paths are in bijection with a class of binary trees by proving that they have identical decompositions.

The number of combinatorial classes that have the Catalan numbers as their counting sequence is overwhelming at times. Many of the object plainly share the same decomposition as the Dyck paths into a first. This does not exhaust the set of classes in bijection by a long shot. It is a classic exercise in bijective combinatorics to try to find bijections between Catalan objects. Famously, Exercise 6.19 in Stanley's *Enumerative Combinatorics II* [Sta99] has grown into the subject of its own book!

Reference: [Sta15]

❐

Exercise 1.2 (A subclass of Dyck paths). Let $\mathscr{C}$ be the subclass of Dyck paths with no contiguous subsequence of $\nearrow\nearrow\nearrow$. To be clear:

$$\mathscr{C}_3 = \{\nearrow\nearrow\searrow\nearrow\searrow\searrow, \nearrow\searrow\nearrow\searrow\nearrow\searrow\}. \tag{1.16}$$

It is straightforward to verify that the classes $\mathscr{C}$ and the set of Motzkin paths, $\mathscr{M}$, both have the same initial terms to their counting sequence: 1, 1, 2, 4, 9, 21, 51, 127, 323, ... Prove that $\mathscr{C} \equiv \mathscr{M}$. ❐

Exercise 1.3. A permutation, σ, is unimodal if there is $k \geq 1$, such that $\sigma_1 < \cdots < \sigma_{k-1} < \sigma_k = n > \sigma_{k+1} > \cdots > \sigma_n$. Find the OGF of the class of unimodal permutations. ❐

Exercise 1.4. Find a size preserving bijective map from Motzkin paths of length n to non-intersecting chord diagrams on n points. (The latter class is all the ways of drawing non-intersecting chords on a circle of n-points. Triangulations are a subclass.) Demonstrate the bijection for $n = 3$. ❐

Exercise 1.5 (Coloured Motzkin paths).

1. Consider a class of modified Motzkin paths such that there are two different kinds of horizontal steps; think of them as being coloured either blue or red. Find the OGF of this class. Is it familiar? Why? Can you find an appropriate bijection?

2. Now consider the class of Motzkin paths where any of the steps can be either red or blue. Prove the bijection between bicoloured Motzkin paths and lattice walks with steps $\{\uparrow, \leftarrow, \searrow, \downarrow, \rightarrow, \nwarrow\}$ that start at the origin and end anywhere, and remain in the first quadrant.

Reference: [CY18]

❐

Exercise 1.6 (Convergence of series). Let $F \in K[[x]]$ be a formal power series and define $\mathrm{ord}(F)$ to be the smallest non-zero power of x in the expansion of F. For example $\mathrm{ord}\left(x^3 \sinh(x)\log(1-x)\right) = \mathrm{ord}(-x^5 - x^6/2 + \dots) = 5$.

For $F, G \in K[[x]]$, define the distance between F and G to be

$$d(F,G) = 2^{-\mathrm{ord}(F-G)},$$

so that F and G are "close" when many of their initial terms agree.

1. Verify that this is a valid distance function.

2. Let $\mathscr{A}^{(1)}, \mathscr{A}^{(2)}, \dots$ be a sequence of combinatorial classes so that $\mathscr{A}^{(i)} \subset \mathscr{A}^{(i+1)}$ with compatible definitions of size. Prove that if $\mathscr{A} = \lim_{i \to \infty} \mathscr{A}^{(i)}$ (as sets) then the following holds for the generating functions of the sets

$$A(x) = \lim_{i \to \infty} A^{(i)}(x).$$

$\square$

Exercise 1.7 (Compositions). A **composition** of an integer n is an increasing sequence of positive integers which sums to n. The class of all compositions is denoted $\mathscr{C}$. For example, $\mathscr{C}_3 = \{1+1+1, 2+1, 1+2, 3\}$.

1. Prove that $\mathscr{C} \equiv (\circ^+)^*$. Deduce $C(x)$ and c_n.

2. Let k be a fixed positive integer. Let $\mathscr{C}^{(k)}$ represent the class of integer compositions where each part is less than k. Show that the generating function for $\mathscr{C}^{(k)}$ is given by

$$C^{(k)}(x) = \frac{1-x}{1-2x+x^k}.$$

3. Find the number of compositions of the positive integer n into odd parts. Express your answer in terms of Fibonacci numbers.

4. Determine an expression for the generating function of compositions such that each part is an integer from set $\mathscr{A}$ and that the size of each part is from set $\mathscr{B}$.

$\square$

Exercise 1.8 (Restricted word classes). Let $\mathcal{W}$ be the combinatorial class of words over the alphabet $\{0, 1, 2\}$ with the restriction that no word has consecutive even numbers. For example, $02 \notin \mathcal{W}_2$. Let w_n be the number of words of length n in $\mathcal{W}_n$.

1. Write a recurrence relation for w_n.

2. Use the recurrence relation to compute w_6.

3. Write a combinatorial grammar to describe the class $\mathcal{W}$.

4. Use the combinatorial grammar to deduce the generating function $W(x)$. Use the series expansion functionality of a computer algebra package to verify w_6.

❏

Exercise 1.9 (Non-plane trees).

1. Consider the ditto operation, which builds copies of substructures:

$$\text{ditto}(\mathcal{C}) = \{(\gamma, \gamma) : \gamma \in \mathcal{C}\}.$$

This operator is often used as an intermediary class when trying to account for symmetries. Prove that ditto is an admissible operation by giving a formula for the generating function of $\text{ditto}(\mathcal{C})$ in terms of $C(x)$.

2. We can use this operator to describe **non-plane** trees. Changing the placement of the children of a node in non-plane tree only changes the representation, it does not create a new tree.

To specify a class of non-plane trees, we do a similar decomposition to the plane tree case: We delete the root node and see what is left over. Instead of an ordered sequence of offspring, we effectively have a set of offspring. We can use the ditto operator to create a functional equation for the generating function.

Let $U(x)$ be the generating function of non-plane binary trees, where size is given by the number of nodes. Prove that $U(x)$ satisfies the following functional equation by interpreting each term combinatorially:

$$U(x) = x + xU(x)^2/2 + xU(x^2)/2.$$

❏

Exercise 1.10. Using the pumping lemma for regular languages prove that the language of palindromes over the alphabet $\Sigma = \{0, 1\}$ is not regular. Is this a context-free language? Either give the grammar or a proof that it is not. ❑

Exercise 1.11. Prove that simple 2-D excursions are not context-free using the pumping lemma. ❑

Exercise 1.12 (Kreweras walks)**.** Consider the quarter-plane walks that start at the origin and remain in the first quadrant using only steps from $\mathscr{S} = \{\nearrow, \leftarrow, \downarrow\}$. These are known as Kreweras walks. Curiously, as a sublanguage of $\mathscr{S}^*$, it is not context-free, however, the generating function is algebraic. The first fact is easy to show, and the second is much harder [BM02]. Find another encoding of these walks that more easily explains the algebraicity of the generating function. (This is hard.)

Reference: [Bud]

❑

Exercise 1.13 (The infamous Gessel walks)**.** A second more complex version of the phenomena of Exercise 1.12 is the lattice model known as **Gessel walks** start at the origin and remain in the first quadrant with step set $\{(-1, 1), (-1, 0), (1, 0), (1, -1)\}$. The algebraicity of the generating function was a subject of vigorous debate for many years and included monetary prizes. See [BR15b] for a summary. The polynomial satisfied by the generating function is very large and incredibly unwieldy! Show that the family of quarter-plane walks, viewed as a language over the step set, is not context-free. This leaves an interesting open question about which combinatorial mechanisms lead to the algebraicity of the generating function.

Reference: (OEIS A135404)[Bud]

❑

Exercise 1.14. Show that the language over $\Sigma = \{a, b\}$, such that each word has the same number of a's and b's, can be described by an algebraic grammar. Show that the language over $\Sigma = \{a, b, c\}$, such that each word has the same number of a's, b's and c's, cannot. Hint: The intersection of a context-free language and a regular language is context-free. ❑

2

Combinatorial Parameters

CONTENTS

2.1	Combinatorial Parameters	36
	2.1.1 Bivariate Generating Functions	36
2.2	What Can We Do with a Bivariate Generating Function?	37
	2.2.1 Higher Moments	38
	2.2.2 Moment Inequalities and Concentration	40
2.3	Deriving Multivariate Generating Functions	41
	2.3.1 Multidimensional Parameters	41
	2.3.2 Inherited Parameters	42
	2.3.3 Marking Substructures	43
2.4	On the Number of Components	46
2.5	Linear Functions of Parameters	46
2.6	Pathlength	47
2.7	Catalytic Parameters and the Kernel Method	49
2.8	Discussion	51
2.9	Problems	51

So far we have seen how to define and count discrete objects using combinatorial calculus. Next we try to understand characteristics of a typical element of the class. We can use multivariable generating functions to compute statistical information about combinatorial properties of the objects. The number of occurrences of a pattern, or the number of atoms of particular type – these are examples of parameters: integer valued functions evaluated on elements in a combinatorial class. We extend the coefficient ring of power series to incorporate more variables to track this additional information. It is important to understand the algebraic context of a multivariate series, particularly when we consider it as analytic objects. For now, we manipulate the series formally using addition, multiplication and quasi inverse, as in the univariate case.

2.1   Combinatorial Parameters

A **parameter** of the combinatorial class $\mathscr{C}$ is a some property of elements of $\mathscr{C}$ that we can compute using a positive integer-valued map $\chi : \mathscr{C} \to \mathbb{N}$. We write this as a pair,

$$(\mathscr{C}, \chi).$$

2.1.1   Bivariate Generating Functions

We define the **bivariate generating function** of $(\mathscr{C}, \chi)$ to be the formal series

$$C(u, x) := \sum_{\gamma \in \mathscr{C}} u^{\chi(\gamma)} x^{|\gamma|}. \tag{2.1}$$

Equivalently:

$$C(u, x) = \sum_{n \geq 0} \left(\sum_{k \geq 0} c_{k,n} u^k \right) x^n, \tag{2.2}$$

where $c_{k,n}$ is the number of objects of size n with parameter value k. We say that u **marks**[1] **the parameter** χ. This series is an element of the set $\mathbb{N}[u][[x]]$, the set of power series in x whose coefficients are polynomials in u with coefficients in $\mathbb{N}$.

Example 2.1 (Binary word parameters). Let $\mathscr{B}$ be the subset of binary words over the alphabet $\{\circ, \bullet\}$ that contain no $\circ\circ$ substring. It is given by the specification

$$\mathscr{B} \equiv (\epsilon + \circ) \times (\bullet \times (\epsilon + \circ))^*.$$

We define the parameter χ to be the number of $\circ$s per word in $\mathscr{B}$. For example, $\chi(\circ \bullet \bullet \circ \bullet) = 2$. We can exhaustively list the words of length at most 3 in $\mathscr{B}$,

$$\mathscr{B} = \{\epsilon, \circ, \bullet, \circ\bullet, \bullet\circ, \bullet\bullet, \circ\bullet\bullet, \bullet\circ\bullet, \bullet\bullet\bullet, \circ\bullet\circ, \bullet\bullet\circ, \dots\}$$

and determine the initial terms of the bivariate generating function for $(\mathscr{B}, \chi)$, where where u marks χ:

$$B(u, x) = 1 + (u + 1)x + (2u + 1)x^2 + \left(u^2 + 3u + 1\right)x^3 + \dots.$$

---

[1]Some authors prefer "u is conjugate to the parameter χ".

Can we determine the generating function explicitly in some closed form? In this chapter we discuss conditions for which there is an analogue to Theorem 1.1, which permits us to compute a closed form for generating functions like $B(u, x)$ systematically:

$$B(u, x) = (1 + ux)\left(\frac{1}{1 - x(1 + ux)}\right). \tag{2.3}$$

◄

2.2 What Can We Do with a Bivariate Generating Function?

Consider a class with a parameter $(\mathscr{C}, \chi)$ and its bivariate generating function $C(u, x)$. First note that evaluating $C(u, x)$ at $u = 1$ returns the counting OGF of the the class. Furthermore, as we have only positive powers of u, we can evaluate u at 0. This annihilates every term in the series except the coefficients of u^0:

$$C(0, x) = [u^0]C(u, x).$$

We obtain the generating function for the subclass of elements whose parameter value is 0 by an evaluation. More generally, a series of the form $[u^k]C(u, x)$ is said to be a **section** of the bivariate series.

Finally, we can say precise things about the distribution of the parameter. In a weighted generating function, each member of the combinatorial class contributes a term weighted by its parameter value. We can write this series as a partial derivative of the bivariate generating function with respect to u, evaluated at $u = 1$:

$$\sum_{\gamma \in \mathscr{C}} \chi(\gamma) x^{|\gamma|} = \frac{\partial}{\partial u} \sum_{\gamma \in \mathscr{C}} u^{\chi(\gamma)} x^{|\gamma|}\bigg|_{u=1}. \tag{2.4}$$

The resulting generating function is called the **cumulative generating function**, and we denote it $C_\chi(x)$:

$$C_\chi(x) := \frac{\partial}{\partial u} C(u, x)\bigg|_{u=1}. \tag{2.5}$$

We use this series to determine the average or expected value of the parameter in a uniform distribution across elements of size n.

Proposition 2.1. *Suppose $\mathscr{C}$ is a combinatorial class and $\chi : \mathscr{C} \to \mathbb{N}$ a parameter. Let $C(u,x)$ be the bivariate generating function of $(\mathscr{C},\chi)$, and $C_\chi(x)$ the cumulative generating function. The expected value of χ for objects of $\mathscr{C}$ of size n satisfies the formula*

$$\mathbb{E}[\chi(\gamma) \mid \gamma \in \mathscr{C}_n] = \frac{[x^n]\frac{\partial}{\partial u}C(u,x)\big|_{u=1}}{[x^n]C(1,x)}.$$

2.2.1   Higher Moments

Let X be the random variable that is the values of χ over $\mathscr{C}_n$, for fixed n. By repeated differentiation of the bivariate generating function $C(u,x)$ with respect to u, followed by the evaluation at $u = 1$, we can determine a formula for the **moments** of X. We recall that the moments of a random variable X are defined as follows:

$$\mathbb{E}[X] = \sum_k \mathbb{P}[X = k]\,k \qquad \text{the expected value of } X$$

$$\mathbb{E}[X^r] = \sum_k \mathbb{P}[X = k]\,k^r \qquad \text{the } r\text{-th moment of } X$$

$$\mathbb{V}[X] = \mathbb{E}[X^2] - \mathbb{E}(X)^2 \qquad \text{the variance of } X.$$

The **factorial moment of order** r is the value

$$\mathbb{E}(X(X-1)(X-2)\ldots(X-r+1)).$$

It can be computed directly from the bivariate generating function.

Proposition 2.2. *Let X be the random variable given by the value of the parameter χ in $\mathscr{C}_n$. Then the factorial moment of order r of X is determined from the bivariate generating function $C(u,x) = \sum_{\gamma \in \mathscr{C}} u^{\chi(\gamma)}x^{|\gamma|}$ by the following formula*

$$\mathbb{E}(X(X-1)\cdots(X-r+1)) = \frac{[x^n]\frac{\partial^r C(u,x)}{\partial u^r}\big|_{u=1}}{[x^n]C(1,x)}.$$

The proof follows directly from the term by term differentiation, after they have been suitably normalized.

We can use the factorial moment of order r to compute the r-moment, by using the linearity of expectation: $\mathbb{E}[X+Y] = \mathbb{E}[X]+\mathbb{E}[Y]$. For example,

$$\mathbb{E}[X(X-1)] = \mathbb{E}[X^2 - X] = \mathbb{E}[X^2] - \mathbb{E}[X].$$

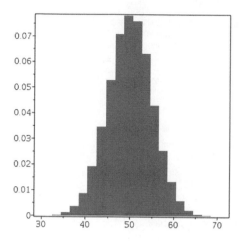

FIGURE 2.1

The Binomial Distribution. The histogram depicts proportion of words with a given value for χ in 100 000 randomly generated binary words of length 100. The values are concentrated around the mean value of 50.

Example 2.2 (Binomial distribution). Consider the class $\mathscr{C}$ of binary words over $\{\circ, \bullet\}$, and the parameter χ counting the number of number $\circ$ per word. Figure 2.1 is a histogram of the value of the parameter for randomly generated words. The distribution of this parameter over all words of length n is called the **binomial distribution**. We can determine an exact formula for the r-th moment from the bivariate generating function. As an application, we can also find the mean and the standard deviation. This is a discrete version of the normal distribution that one frequently encounters. We can compute higher moments from the bivariate generating function, where u marks the occurrences of $\circ$. The bivariate generating function is

$$C(u,x) = \sum_{n\geq 0}\left(\sum_{k=0}^{n}\binom{n}{k}u^k\right)x^n = \sum_{n\geq 0}(1+u)^n x^n = \frac{1}{1-x(1+u)}.$$

We can explicitly determine the r-th derivative with respect to u from this expression:

$$\left(\frac{\partial^r}{\partial u^r}\frac{1}{1-x(1+u)}\right)\bigg|_{u=1} = \frac{x^r\, r!}{(1-x(1+u))^{r+1}}\bigg|_{u=1} = \frac{x^r\, r!}{(1-2x)^{r+1}}.$$

Using Newton's generalized binomial theorem,

$$[x^n]\frac{x^r r!}{(1-2x)^{r+1}} = 2^{n-r} r! \binom{n}{r}.$$

Thus the mean is computed to be $\frac{2^{n-1} 1! \binom{n}{1}}{2^n} = \frac{n}{2}$ (as expected!), and the variance is

$$E[X^2] - E[X]^2 = E[X(X-1)] + E[X] - E[X]^2$$

$$= \frac{n(n-1)}{4} + \frac{n}{2} - \frac{n^2}{4} = \frac{n}{4}.$$

Thus the standard deviation is $\sqrt{E[X^2] - E[X]^2} = \sqrt{n}/2$, which grows slowly compared to the mean. We conclude that the distribution is quite concentrated around the mean.   ◀

When we can compute or estimate moments, we can get a first understanding as to how the parameter is distributed.

2.2.2   Moment Inequalities and Concentration

The following result tells us that the probability of a random variate being larger than the mean must decay and that this decay is governed by the standard deviation.

Proposition 2.3 (Markov-Chebyshev inequalities). *Let X be a non-negative random variable and Y any real variable. One has for any $t > 0$ that*

$$\mathbb{P}(X \geq t\mathbb{E}[X]) \leq \frac{1}{t} \qquad \textit{Markov inequality}$$

$$\mathbb{P}(|Y - \mathbb{E}[X]| \geq t\sigma(Y)) \leq \frac{1}{t^2} \qquad \textit{Chebyshev inequality}$$

We use this to determine how distributions concentrate around their means. In the context of parameters on classes, we set X_n to be the random variable that is the value of the parameter χ applied to elements in the class $\mathscr{C}_n$.

Proposition 2.4 (Concentration of distribution). *Consider a family of random variables $\{X_n\}$. If the mean μ_n and the standard deviation σ_n of X_n satisfy*

$$\lim_{n \to \infty} \frac{\sigma_n}{\mu_n} = 0,$$

then given any $\epsilon > 0$

$$\lim_{n \to \infty} \mathbb{P}\left(1 - \epsilon \leq \frac{X_n}{\mu_n} \leq 1 + \epsilon\right) = 1.$$

That is to say the distribution of the X_n is concentrated around the mean.

In short, provided the standard deviation grows slowly compared to the mean, the values of the parameters become increasingly close to the mean. This is useful information to know when trying to understand the large-scale behaviour of a combinatorial class.

2.3 Deriving Multivariate Generating Functions

Next we describe how to derive bivariate (and ultimately multivariate) generating functions for the classes specified by the admissible operators and grammars we saw in the last chapter.

2.3.1 Multidimensional Parameters

We can consider multiple parameters at once with the same techniques. Consider a class $\mathscr{C}$. A **multidimensional parameter** $\chi = (\chi_1, \ldots, \chi_d)$ is a function from $\mathscr{C}$ to $\mathbb{N}^d$.

The **multivariate generating function** (MGF) of $\mathscr{C}$ marking multiparameter χ by u is the series:

$$C(u_1, \ldots, u_d, x) := \sum_{\gamma \in \mathscr{C}} u_1^{\chi_1(\gamma)} \ldots u_d^{\chi_d(\gamma)} x^{|\gamma|}$$

$$= \sum_{\substack{n \geq 0 \\ \mathbf{k} \in \mathbb{N}^d}} \text{card}\{\gamma \mid |\gamma| = n, \chi_1(\gamma) = k_1, \ldots, \chi_d(\gamma) = k_d\} \mathbf{u}^{\mathbf{k}} x^n.$$

Here we use the natural multinomial exponentiation:

$$\mathbf{u}^{\mathbf{k}} = u_1^{k_1} \ldots u_d^{k_d}.$$

This is an element of $\mathbb{N}[u_1, \ldots, u_d][[x]]$.

In analogy to the univariate case, we recover the single variable ordinary generating function by setting each of the $u_i = 1$. Setting $u_i = 0$ for any $1 \leq i \leq d$ determines the generating function of the subclass satisfying the condition that $\chi_i(\gamma) = 0$.

2.3.2   Inherited Parameters

When an combinatorial class is given by a grammar, or simply some admissible construction, a parameter may be computed based on parameter values for the object's composite parts. The size of a structure is an example of such a parameter. In such a case, we say that a parameter value is **inherited**, and often we can translate the combinatorial information directly into generating function information. We make the definition precise below.

Let $(\mathscr{A}, \xi), (\mathscr{B}, \zeta)$ and $(\mathscr{C}, \chi)$ be three classes with parameters of the same dimension d.

Combinatorial Sum

Suppose that $\mathscr{C} = \mathscr{A} + \mathscr{B}$. The parameter χ is inherited from ξ and ζ if and only if

$$\chi(\gamma) = \begin{cases} \xi(\gamma) & \gamma \in \mathscr{A} \\ \zeta(\gamma) & \gamma \in \mathscr{B}. \end{cases}$$

We denote this relationship by $(\mathscr{C}, \chi) = (\mathscr{A}, \xi) + (\mathscr{B}, \zeta)$.

Cartesian Product

For a class $\mathscr{C}$ defined as $\mathscr{C} = \mathscr{A} \times \mathscr{B}$, the parameter χ is inherited from ξ and ζ if and only if

$$\chi(\alpha, \beta) = \xi(\alpha) + \zeta(\beta).$$

Accordingly we may write $(\mathscr{C}, \chi) = (\mathscr{A}, \xi) \times (\mathscr{B}, \zeta)$.

Sequence

If the combinatorial class $\mathscr{C} = \text{SEQ}(\mathscr{A})$, then the parameter χ is inherited from ξ if and only if

$$\chi(\beta_1, \ldots, \beta_r) = \xi(\beta_1) + \cdots + \xi(\beta_r)$$

We write $(\mathscr{C}, \chi) = \text{SEQ}(\mathscr{C}, \xi)$, and remark that inherited parameters of restricted sequences and powers is similar.

We may take the convention that $\chi_{d+1}(\gamma) = |\gamma|$, and then mark the set of parameters by $\mathbf{x} = x_1, \ldots, x_{d+1}$. In this case we use the shorthand notation

$$C(\mathbf{x}) = \sum_{n \in \mathbb{N}^{d+1}} c_n \mathbf{x}^n,$$

an element of $\mathbb{N}[x_1 \ldots, x_d][[x_{d+1}]]$.

Theorem 2.5 (Inherited parameter generating functions). *Suppose that $\mathscr{C}$ is a class with inherited parameter χ, then the following generating function relations are true:*

Combinatorial Sum	$(\mathscr{C},\chi)=(\mathscr{A},\xi)+(\mathscr{B},\zeta)$	$C(\mathbf{x})=A(\mathbf{x})+B(\mathbf{x})$
Cartesian Product	$(\mathscr{C},\chi)=(\mathscr{A},\xi)\times(\mathscr{B},\zeta)$	$C(\mathbf{x})=A(\mathbf{x})B(\mathbf{x})$
Sequence	$(\mathscr{C},\chi)=\textsc{Seq}(\mathscr{A},\xi)$	$C(\mathbf{x})=\dfrac{1}{1-A(\mathbf{x})}$

The proof is done case by case from direct algebraic manipulation of the generating functions. The base cases are often the atomic and epsilon classes. Their multivariable generating functions should be clear from the definition of the problem.

Example 2.3. We apply Theorem 2.5 to Example 2.1. Recall that $\chi(w)$ is the number of o in w. Thus $\chi(\mathrm{o})=1$ and $\chi(\epsilon)=\chi(\bullet)=0$. The value of χ for other objects follows from inheritance:

$$
\begin{array}{ccccc}
\mathscr{B} & \equiv & (\epsilon+\mathrm{o}) & \times & (\bullet(\epsilon+\mathrm{o}))^* \\
\downarrow & & \downarrow & & \downarrow \\
B(u,x) & = & (1+ux) & \cdot & \frac{1}{1-x(1+ux)}
\end{array}
$$

◀

2.3.3 Marking Substructures

Next we formalize this strategy to use inherited parameters to track the number of occurrences of a particular substructure in an object. A systematic technique to do this is to tag objects that we want to count with an epsilon class (say μ). We define $\chi(\mu)=1$ and then propagate the parameter value up using inherited parameters.

Example 2.4 (The number of leaves in a general plane tree). Let us mark the leaves in a plane tree with such an epsilon class. The nodes in the tree are atomic elements, o, then the specification of plane trees where the epsilon tag μ marks leaves is:

$$
\mathscr{T}=\mathrm{o}\times\textsc{Seq}_{\geq 1}(\mathscr{T})+\mu\times\mathrm{o}.
$$

The tree representation of a small example is then:

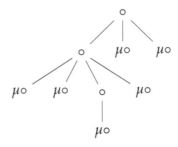

The number of μ-tags is precisely the number of leaves, in this case 6. Given this, we can translate this to a system of functional equations for the generating functions. Note, we define $Z(u,x)$ to be the generating function for the atomic class $\{\circ\}$, and $M(u,x)$ to be the generating function for epsilon class $\{\mu\}$. By Theorem 2.5,

$$T(u,x) = Z(u,x)\frac{T(u,x)}{1 - T(u,x)} + M(u,x)Z(u,x)$$

$$Z(u,x) = x, \qquad \text{since } \chi(\circ) = 0$$

$$M(u,x) = u$$

$$\implies T(u,x) = z\frac{T(u,x)}{1 - T(u,x)} + zu.$$

We solve for $T(u,x)$:

$$T(u,x) = \frac{xu - x + 1 - \sqrt{u^2x^2 - 2ux^2 - 2xu + x^2 - 2x + 1}}{2}$$

$$= xu + ux^2 + (u^2 + u)x^3 + \ldots$$

With the explicit generating function we can ask questions about the distribution of the parameter, which we leave to the exercises. ◄

We consider now a second example, which illustrates to some very general phenomena.

Example 2.5 (The number of summands in a composition). A **composition** of a positive integer n is an ordered list of positive integers that sums to n. For example, the compositions of 3 are $(1,1,1),(1,2),(2,1),(3)$. We can view a composition as a sequence of non-empty blocks of atoms (summands):

$$\mathscr{C}_3 = \{\circ | \circ | \circ, \quad \circ | \circ \circ, \quad \circ \circ | \circ, \quad \circ \circ \circ\}.$$

This is given by the specification

$$\mathscr{C} = \text{SEQ}(\mathscr{B}) \quad \mathscr{B} = \text{SEQ}_{\geq 1}(\circ).$$

The generating function for compositions is $\frac{2x}{1-2x}$. We can determine the mean number of summands in a composition. Let $\chi(\gamma)$ be the number of summands in a composition. For example,

$$\chi(\circ|\circ|\circ\circ\circ\circ\circ|\circ\circ) = 4.$$

To count the number of summands we tag each part with μ, an epsilon class:

$$\mathscr{C} = \text{SEQ}(\mathscr{B}) \quad \mathscr{B} = \mu \times \text{SEQ}_{\geq 1}(\circ).$$

This is an inherited parameter

$$(C, \chi) = \text{SEQ}((B, \chi)) \quad (B, \chi) = (\mu, \chi) \times \text{SEQ}((\circ, \chi)) \quad \chi(\circ) = 0 \quad \chi(\mu) = 1.$$

In the modified grammar a composition is represented as follows:

$$\chi(\mu\circ|\mu\circ|\mu\circ\circ\circ\circ\circ|\mu\circ\circ) = 4,$$

and the number of summands is equal to the number of μ.

We translate the specification using Theorem 2.5 into a system for the generating functions:

$$C(u, x) = \frac{1}{1 - C(u, x)} \qquad\qquad Z(u, x) = x$$

$$B(u, x) = M(u, x)\frac{1}{1 - Z(u, x)} \qquad\qquad M(u, x) = u.$$

This is solved:

$$C(x, u) = \frac{1}{1 - u\frac{x}{1-x}}.$$

We can find the cumulative generating function and deduce the mean number of summands:

$$C_\chi(x) = \frac{\partial}{\partial u}C(x, u)\Big|_{u=1} = \frac{x(1 - x)}{(1 - 2x)^2}$$

$$= \frac{1/4}{(1 - 2x)^2} - \frac{1}{4}.$$

Hence the mean number of summands in a composition of size n is

$$\frac{[x^n]C_\chi(x)}{[x^n]C(1, x)} = \frac{2^{n-2}(n + 1)}{2^{n-1}} = \frac{n + 1}{2}.$$

Proposition 2.6. *The mean number of summands in a composition of* n *is* $\frac{n+1}{2}$

Additional parameters of compositions are considered in the exercises.   ◄

2.4   On the Number of Components

As we can see from the composition example, the number of components in a sequence is a parameter that is straightforward to analyze by the following process. Suppose the class $\mathscr{C}$ is composed of objects of type $\mathscr{A}$. We modify the specification with a μ tag of size 0 to each $\mathscr{A}$ object:

$$\mathscr{C} = \text{Seq}(\mathscr{A}) \longrightarrow \mathscr{C} = \text{Seq}(\mu \times \mathscr{A}). \tag{2.6}$$

Then we deduce that

$$C_\chi(x) = \left.\frac{\partial}{\partial u}C(u,x)\right|_{u=1} = \frac{A(x)}{(1-A(x))^2} = C(x)^2 \cdot A(x),$$

where $A(x)$ and $C(x)$ are the univariate counting generating functions of the classes $\mathscr{A}$ and $\mathscr{C}$, respectively. Thus, if X is the random variable that is the number of components in an object in $\mathscr{C}_n$, then the expected value of X is

$$\mathbb{E}(X) = \frac{[x^n]C_\chi(x)}{[x^n]C(x)} = \frac{[x^n]C(x)^2 \cdot A(x)}{[x^n]C(x)}.$$

2.5   Linear Functions of Parameters

Some parameters are best defined in terms of other parameters in a linear way. We refer to such a parameter as an **additive parameter**. Suppose that χ and ξ are two parameters defined in the class $\mathscr{C}$ that are related in the following way: $\chi(\gamma) = a\xi(\gamma) + b|\gamma|$ for positive integer constants a, b.

Then

$$C_\chi(u,x) := \sum_{\gamma \in \mathscr{C}} x^{|\gamma|} u^{\chi\gamma}$$

$$= \sum_{\gamma \in \mathscr{C}} x^{|\gamma|} u^{a\xi(\gamma)+b|\gamma|}$$

$$= \sum_{\gamma \in \mathscr{C}} (xu^b)^{|\gamma|} u^{a\xi(\gamma)}$$

$$= C_\xi(u^a, u^b x).$$

We can summarize this idea:

A linear transformation of parameters induces a monomial substitution on the corresponding marking variables.

2.6 Pathlength

A key parameter that arises as a linear function is the **mean pathlength of a rooted tree**. The length of a path in a tree is the number of edges between the the endpoints. The total pathlength of a tree τ is the sum of all the distances from the vertices in the tree to the root: $\chi(\tau) = \sum_{v \in V(\tau)} d(v, \text{root})$. We take the convention the the distance from the root to itself is zero. Figure 2.2 gives an example.

Consider the example of the tree τ in Figure 2.2. Consider any vertex in the subtree τ' rooted at the grey vertex. The distance between that vertex to the root of τ is exactly one more than the distance to the grey vertex plus one, since the path must pass through the root of τ'. Thus, the sum

$$\sum_{v \in V(\tau')} d(v, \text{root}(\tau)) = \sum_{v \in V(\tau')} (d(v, \text{root}(\tau')) + 1) = \chi(\tau') + |\tau'|.$$

Summing over all of the subtrees gives the formula:

$$\chi(\tau) = \sum_{\tau' < \tau} \chi(\tau') + |\tau'| \tag{2.7}$$

where $\tau' < \tau$ indicates that τ' is a tree rooted at a direct descendant of the root of τ.

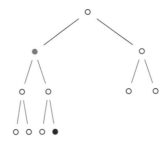

FIGURE 2.2
A rooted tree τ. The distance from the node $\bullet$ to the root is 3. The total pathlength of this tree is 16.

Example 2.6 (The average pathlength of a binary tree). Let us consider the case of binary trees defined by the specification

$$\mathcal{B} = \mathcal{Z} + \mathcal{Z} \cdot \mathcal{B}^2.$$

We can determine a functional equation for the cumulative generating function:

$$B(u,x) := \sum_{\tau \in \mathcal{B}} u^{\chi(\tau)} x^{|\tau|} \tag{2.8}$$

$$= x + x \sum_{\tau = \circ \times \tau_1 \times \tau_2} u^{\chi(\tau_1)} u^{\chi(\tau_2)} (ux)^{|\tau_1| + |\tau_2|} \qquad \text{using Eq. (2.7)}$$

$$\tag{2.9}$$

$$= x + x \sum_{\tau_1 \in \mathcal{B}} u^{\chi(\tau_1)} (ux)^{|\tau_1|} \sum_{\tau_2 \in \mathcal{B}} u^{\chi(\tau_2)} (ux)^{|\tau_2|} \tag{2.10}$$

$$= x + x B(u, ux)^2. \tag{2.11}$$

This equation is difficult to solve, however, we can use it to solve for $B_\chi(x)$:

$$B_\chi(x) = \frac{\partial}{\partial u} x B(u, ux)^2 \Big|_{u=1} \tag{2.12}$$

$$= [2xB(u, ux)(B_u(u, ux) + xB_x(u, ux))]_{u=1} \tag{2.13}$$

$$= 2xB(x)\Big(B_\chi(x) + xB'(x)\Big). \tag{2.14}$$

Since we know $B(x) = \frac{1 - \sqrt{-4x^2 + 1}}{2x}$, this is enough information to solve for $B_\chi(x)$:

$$B_\chi(x) = \frac{2x^2 + \sqrt{-4x^2 + 1} - 1}{4x^3 - x} = 2x^3 + 12x^5 + 58x^7 + O\left(x^9\right).$$

From this expression we can determine that the expected sum

$$\frac{[x^n]B_\chi(x)}{[x^n]B(x)} = \frac{\sqrt{\pi n^3}}{2}.$$

Hence the average distance of a node to the root is roughly $\sqrt{n}$.   ◂

Similarly, we can prove that for a general class of Ω-trees the bivariate generating function $T(u, x)$ for the pathlength satisfies

$$T(u, x) = x\Omega(T(u, ux)).$$

2.7   Catalytic Parameters and the Kernel Method

Parameters sometimes appear in order to facilitate a recurrence or a decomposition. This is common, for example, in the enumeration of lattice paths with boundary conditions. Parameters that are introduced with the intent of facilitating enumeration are called **catalytic parameters**, and they are marked by **catalytic variables**.

Example 2.7 (Dyck paths). We present another method for counting simple excursions in one dimension by considering the larger class of walks with ↗ and ↘ steps, that always remain at a nonnegative height, but are no longer bound to return to the axis. We illustrate how to use the catalytic parameter that is the height at the end of the walk. This information permits us to decompose the walks as follows:

A walk is either the empty walk, or a walk plus a valid step. A down step is not valid to append to a walk at height 0, wherein we have used the height information.

We mark the end height with the variable u and consider the bivariate generating function for this parameter:

$$C(u, x) := \sum_{n \geq 0}\left(\sum_{k \geq 0} c_{n,k} u^k\right) x^n,$$

where $c_{n,k}$ is the number of walks of length n ending at height k. Since the walks cannot end below the axis, we remark that $C(u,x) \in \mathbb{N}[u][[x]]$. The **excursions** are the walks that end at height 0. The generating function of excursions is the section $[u^0]C(u,x)$, which is equal to $C(0,x)$ here. The decomposition translates into a generating function equation:

$$C(u,x) = 1 + xuC(u,x) + \frac{x}{u}(C(u,x) - C(0,x)). \qquad (2.15)$$

This is a different kind of functional equation than we have encountered so far, as there are essentially two unknowns, $C(u,x)$ and $C(0,x)$, even though they are related to each other. We solve it with a method called the **kernel method**, which has several different forms.

Before solving, however, we do a quick check to see that indeed there is a series solution to this equation. We can generate terms by iterating

$$C^{[k+1]}(u,x) = 1 + xuC^{[k]}(u,x) + \frac{x}{u}\left(C^{[k]}(u,x) - C^{[k]}(0,x)\right). \qquad (2.16)$$

starting with $C^{[0]}(u,x) = 0$. We can see that at each iteration additional terms are stabilized, and so the limit $\lim_{k \to \infty} C^{[k+1]}(u,x)$ exists and is the solution to the equation.

To solve this, the first step is to collect the $C(u,x)$ on the left-hand side, and write what is called the **kernel version** of the equation:

$$C(u,x)\left(1 - xu - \frac{x}{u}\right) = 1 - \frac{x}{u}C(0,x). \qquad (2.17)$$

Next, we find a function $u = U(x)$ that satisfies $1 - xU(x) - \frac{x}{U(x)} = 0$. Such a function must be invertible, and the substitution into $C(u,x)$ must make sense. (Not all series substitutions result in series). In this case, we find such a series by solving a quadratic equation: $U(x) = 1/2 \frac{1-\sqrt{-4x^2+1}}{x}$ We verify that $U(x) = 1/2 \frac{1-\sqrt{-4x^2+1}}{x}$ has a zero constant term, and hence can be substituted into $C(u,x)$. Under such a substitution, Eq. (2.17) becomes

$$0 = 1 - \frac{x}{U(x)}C(0,x) \implies C(0,x) = \frac{1}{x}U(x). \qquad (2.18)$$

Once we have $C(0,x)$, we determine $C(u,x)$ from Eq. (2.17).   ◄

2.8 Discussion

One strong motivator for parameter analysis is the average case analysis of algorithms. Tree parameters are very well studied as they are heavily implicated in the analysis of algorithms. The height of a tree is the maximal distance between the root and a leaf. It is difficult to convert the easy combinatorial description into an analytic. See [FGOR93] as a starting point for more details.

There are numerous general results on the distribution of the parameter number of components. We only described sequences here, but there is a well-developed theory to treat other collection structures like sets and cycles. See [DGPR10] for more details.

Banderier and Flajolet use the kernel method to study 1-D walks. They are able to determine explicit generating function expressions, and asymptotic expansions for several types of walks including excursions, and walks that end at any position.

There are many functional equations, particularly related to lattice walks in bounded regions which can be solved using the kernel method. Temperley in 1956 published a similar strategy to study column convex polytopes. In the 1960s, Tutte studied planar maps. His formulas for triangulations use a catalytic parameter, and he can deduce very elegant enumeration formulas. Several families of permutations and lattice walks have used this strategy in concert with kernel method techniques.

What can be said about the nature of the resulting generating functions? In the case of a single catalytic variable, Bousquet-Mélou and Jehanne determine conditions which ensure algebraicity of the generating functions.

2.9 Problems

Exercise 2.1. Let χ be the parameter defined on Motzkin paths that is the number of horizontal steps. Determine the mean value for walks of length n. Is this parameter concentrated about the mean? ❐

Exercise 2.2. Write a script to determine the r-th moment from the r-th factorial moment. ❐

Exercise 2.3. Prove Theorem 2.5.   ❐

Exercise 2.4. Compare the average pathlength parameter for various tree classes. Try to understand how it changes with Ω.   ❐

Exercise 2.5 (Degree of the root). Determine the average degree of the root of a plane tree. Start with a specification that distinguishes between the main tree whose root will be marked, denoted $\mathscr{T}^{\mu}$, and subsequent trees whose roots are not marked:

$$\mathscr{T}^{\mu} = \mu \times \circ \times \text{Seq}(\mathscr{T})$$
$$\mathscr{T} = \circ \times \text{Seq}(\mathscr{T}).$$

Show that the expected degree of the root of a tree on n nodes is $3\frac{n-1}{n+1}$. Is this concentrated around the mean? Hint! You may find it helpful to use the generating relation $T(x) = \frac{x}{1-T(x)}$ to simplify your expressions at some point.   ❐

Exercise 2.6 (On the number of rs in a composition). Let χ_r be the number of summands equal to r (an r-summand) in a composition. For example, $\chi_2(1+2+3+4+2+2+1) = 3$. Determine the generating function for compositions without any summand equal to r. Prove that the mean number of r-summands in a composition of n is

$$\frac{n}{2^{r+1}} + o(1)$$

with standard deviation $\sqrt{n}$. Notice that this means that the number of r-summands is decaying exponentially with r. There are about $n/4$ 1s, $n/8$ 2s and so on. This gives a very clear picture of what the typical large composition looks like.

Reference: [FS09, Chp III]

                                                                       ❐

Exercise 2.7. Let $\chi(w)$ be the area under a Dyck path above the x-axis. For example $\chi(\diagup\diagdown\diagup\diagdown) = 2$. Use a factoring decomposition, write χ as a recursive additive parameter. Determine the average area under a Dyck path that starts at $(0,0)$ and ends at $(2n,0)$.   ❐

Exercise 2.8. Let $\mathscr{B}$ be the class of binary words over $\{\circ, \bullet\}$ with no $\circ\circ$, and define the parameter χ, which counts the number of $\circ$. Make a histogram of values of χ for 10 000 randomly generated values to compare to the binary distribution. Explicitly compute the mean value and standard deviation of χ over $\mathscr{B}_n$.   ❐

Exercise 2.9 (Tracking parameters of walks). Consider Dyck paths to be paths starting at $(0,0)$ and ending at $(2n,0)$ taking steps $(1,1)$ and $(1,-1)$ staying above the line $y = 0$. A contact in such a path is a vertex lying in the line $y = 0$.

1. Find the bivariate generating function counting Dyck paths by their half-length (half the number of edges), where u marks the number of contacts.

2. Use a clever factorisation and Lagrange inversion to find as nice an expression as you can for the number of paths of length $2n$ with k contacts.

$\square$

3

Derived and Transcendental Classes

CONTENTS

3.1   The Diagonal of a Multivariable Series 56
    3.1.1   The Ring of Formal Laurent Series 58
    3.1.2   Basic Manipulations 59
    3.1.3   Algebraic Functions Are Diagonals 60
    3.1.4   Excursion Generating Functions 61
3.2   The Reflection Principle 65
    3.2.1   A One-dimensional Reflection 66
    3.2.2   A Two-dimensional Reflection 68
3.3   General Finite Reflection Groups 70
    3.3.1   A Root Systems Primer 70
    3.3.2   Enumerating Reflectable Walks 72
    3.3.3   A Non-simple Example: Walks in A_2 73
3.4   Discussion 75
3.5   Problems 76

Many interesting combinatorial classes simply cannot be described using an algebraic grammar. Provably so! Borrowing number theoretic terminology we call a class that does not have an algebraic generating function a **transcendental combinatorial class**. There are other kinds of exploitable structures possible, such as being a natural subclass of an algebraic or a S-regular class. In this chapter, we consider a particular kind of subclass that selects elements based on the values of certain parameters. We determine generating functions using a map from multivariable power series to univariate power series, which extracts the terms that correspond to these parameter values. This map is known as the diagonal operator, and it is extremely well studied in number theory, algebra and physics. Here we will illustrate its utility in combinatorics. Indeed, some of the combinatorial techniques described in this chapter are over a hundred years old, dating back to André in the late-19th century in his solution to the two-candidate ballot problem. It is more recent that we have systematic and effective methods to extract subseries from a generating function with some facility. We will investigate subclasses of elements with a particular value for a given inherited parameter. Because they are subclasses, and because the resulting generating functions satisfy a linear differential equation, we call the family of classes studied in this chapter **derived classes**.

3.1 The Diagonal of a Multivariable Series

We begin with multidimensional series. Fix d, the dimension, and recall some standard vector notation:

$$\mathbf{x} := x_1, \ldots, x_d; \quad \mathbf{x}^{\mathbf{n}} := x_1^{n_1} \ldots x_d^{n_d}.$$

Throughout this section let $F(\mathbf{x})$ denote the series:

$$F(\mathbf{x}) = \sum_{(n_1,\ldots,n_d) \in \mathbb{N}^d} f(n_1,\ldots,n_d) x_1^{n_1} \ldots x_d^{n_d} = \sum_{\mathbf{n} \in \mathbb{N}^d} f(\mathbf{n}) \mathbf{x}^{\mathbf{n}}.$$

Here series act as an encoding of a multidimensional array. More formally, we can identify it as an element of $K[[x_1]]\ldots[[x_d]]$. Note, this is a larger ring than $K[x_1,\ldots,x_{d-1}][[x_d]]$, the ring of power series in x_d with polynomial coefficients considered in the last chapter. We say that $F(\mathbf{x})$ is **combinatorial** if $f(\mathbf{n}) \geq 0$ for all $\mathbf{n} \in \mathbb{N}^d$.

The **constant term** of $F(\mathbf{x})$, is defined using a map

$$\mathrm{CT} : K[[x_1]]\ldots[[x_d]] \to K$$

defined as:

$$\mathrm{CT}\, F(\mathbf{x}) = \mathrm{CT} \sum_{\mathbf{n} \in \mathbb{N}^d} f(\mathbf{n}) \mathbf{x}^{\mathbf{n}} := f(0,0,\ldots,0).$$

We denote the constant term with respect to a subset of variables using subscripts. Given a function, if it is clear which series is meant, we can apply CT to the function. We will say more about this important but subtle point in a moment.

The **central diagonal** is a map

$$\Delta : K[[x_1]][[x_2]]\ldots[[x_d]] \to K[[x_d]].$$

defined as:

$$\Delta F(\mathbf{x}) = \Delta \sum_{\mathbf{n} \in \mathbb{N}^d} f(\mathbf{n})\mathbf{x}^{\mathbf{n}} := \sum_{n \geq 0} f(n, n, \dots, n) x_d^n. \tag{3.1}$$

As a default, we use the convention to express the resulting univariate series in the last listed variable. A diagonal with respect to a subset of the variables is specified by a subscript. In this case, we consider the remaining variables to be part of the coefficient ring. For example:

$$\Delta(x^2yz + 3xyz + 7xyz^2 + 5x^2y^2z^2) = 3z + 5z^2,$$

yet the diagonal with respect to x and y has z in the coefficient ring:

$$\Delta_{x,y}(x^2yz + 3xyz + 7xyz^2 + 5x^2y^2z^2) = (3z + 7z^2)y + 5z^2y^2.$$

Finally, we define **diagonal along the ray** $\mathbf{r} = (r_1, r_2, \dots, r_d)$:

$$\Delta^{\mathbf{r}} F(\mathbf{x}) = \Delta^{\mathbf{r}} \sum_{\mathbf{n} \in \mathbb{N}^d} f(\mathbf{n})\mathbf{x}^{\mathbf{n}} := \sum_{n \geq 0} f(nr_1, nr_2, \dots, nr_d) x_d^n.$$

If a function has a Taylor expansion around the origin, then using this series, the constant term and the diagonal are well defined. We can also define diagonals and constant terms for series in larger rings, but first consider an example.

Example 3.1 (Multinomials). The rational $\frac{1}{1-x-y}$ has a straightforward Taylor expansion around the origin, from which we deduce an exact expression for the coefficients of its central diagonal:

$$\Delta\frac{1}{1-x-y} = \Delta \sum_{n \geq 0}(x+y)^n = \Delta \sum_{\ell \geq 0}\sum_{k \geq 0}\binom{\ell+k}{k}x^k y^\ell = \sum_{n \geq 0}\binom{2n}{n}y^n.$$

Indeed, we can determine an exact expression for the diagonal along any ray:

$$\Delta^{(r,s)}\frac{1}{1-x-y} = \sum_{n \geq 0}\binom{rn+sn}{rn}y^n.$$

This example generalizes naturally to arbitrary dimension, using multinomials:

$$\Delta^{\mathbf{r}}\frac{1}{1-(x_1+\cdots+x_d)} = \sum_{n \geq 0}\binom{n(r_1+\cdots+r_d)}{nr_1,\dots,nr_d}x_d^n.$$

◀

These examples are the archetypical diagonals of rational functions. We quantify this as follows. A **(multiple) binomial sum** over a field K is an element in the smallest K-algebra of sequences which contains:

- the sequence $a(0) = 1, a(n) = 0, n > 0$;

- geometric sequences;

- and binomial coefficient sequences.

The diagonal of a rational function is the generating function of a binomial sum.

Given this form it may be unsurprising to learn that the coefficient sequence of a diagonal of a rational function satisfies a finite linear recurrence with polynomial coefficients. That is, it is a **P-recursive sequence**. For example the sequence $b(n) = \binom{2n}{n}$ satisfies the recurrence $(n+1)^2 b(n+1) - 2(n+1)b(n) = 0$. Consequently, a diagonal of a rational function satisfies a linear differential equation with polynomial coefficients. Such functions are said to be **D-finite** and are important in combinatorics and special function theory. We study them in detail in Chapter 5.

We express generating functions in terms of diagonals in this chapter to offer insight about the structure, and because in many cases we can estimate the coefficients. In Part II, we give methods to analyze the asymptotic behaviour of the coefficients of diagonals of rational functions. This is but one motivation to write generating functions in this form.

3.1.1   The Ring of Formal Laurent Series

Not every function in our interest has a Taylor expansion around the origin. Many examples use Laurent series and iterated Laurent series. The ring of **Laurent polynomials** in x over a ring R is denoted $R[x, x^{-1}]$. This is the set of linear combinations of positive and negative powers of x with coefficients in R. A formal **Laurent series** in one variable is a series of the form $\sum_{n \in \mathbb{Z}} f_n x^n$, such that the number f_n with negative indices is finite. The ring of formal Laurent series over ring R is denoted $R((x))$. When $R = K$ is a field then $K((x))$ is a field and may alternatively be obtained as the field of fractions of the integral domain $K[[x]]$.

These formalities are important in order to ensure that coefficient extraction in the multivariable series case is well defined. The following example illustrates the key issue.

Example 3.2. Consider the rational function $\frac{1}{y-x} \in \mathbb{Z}(x,y)$. If we consider this first as a series in x, then we have

$$\frac{1}{y-x} = \frac{1}{y}\frac{1}{1-\frac{x}{y}} = \sum_{n\geq 0}\frac{x^n}{y^{n+1}}.$$

This is an element of $\mathbb{Z}[y^{-1}][[x]]$. Similarly, if we view this first as a series in y, then

$$\frac{1}{y-x} = -\sum_{n\geq 0}\frac{y^n}{x^{n+1}} \in \mathbb{Z}[x^{-1}][[y]].$$

Formally within their own ring, each series is an inverse to the element $\frac{1}{y-x}$, but $[x^k y^\ell]\frac{1}{y-x}$ is not well-defined without a precision of which of these two rings is intended. ◄

The problem of the ambiguity of the series expansion for $F(\mathbf{x})$ is resolved by fixing an order to the indeterminates $x_1,\ldots,x_d$ a priori, and viewing the series as living in $K((x_1))((x_2))\ldots((x_d))$. In this case, we say it is an **iterated Laurent Series**.

3.1.2 Basic Manipulations

We can relate the constant term expression and diagonals directly using formal series manipulations. In many combinatorial applications, the generating functions are power series in x_d with coefficients in an iterated ring of Laurent polynomials, such as

$$K[x_1,x_1^{-1}]\ldots[x_d,x_d^{-1}][[x_{d+1}]] \equiv K[x_1,x_1^{-1},\ldots,x_d,x_d^{-1}][[x_{d+1}]].$$

Coefficient extraction is well-defined here, starting with x_d.

Proposition 3.1. *Suppose that* $F(\mathbf{x}) \in K[x_1,x_1^{-1}]\ldots[x_d,x_d^{-1}][[x_{d+1}]]$. *Then the following equivalence holds:*

$$\mathrm{CT}\, F(\mathbf{x}) = \Delta F\left(\frac{1}{x_1},\frac{1}{x_2},\ldots,\frac{1}{x_d},x_1 x_2 \ldots x_{d+1}\right).$$

Proof. This is proved by direct formal series manipulation. The substitution is well-defined in this ring:

$$F\left(\frac{1}{x_1}, \frac{1}{x_2}, \ldots, \frac{1}{x_d}, x_1 x_2 \cdots x_{d+1}\right)$$

$$= \sum_{\mathbf{n} \in \mathbb{Z}^d, n_d \geq 0} f(\mathbf{n}) x_1^{-n_1} \cdots x_d^{-n_d} (x_1 \ldots x_{d+1})^{n_{d+1}}$$

$$= \sum_{\mathbf{n} \in \mathbb{Z}^d, n_d \geq 0} f(\mathbf{n}) x_1^{n_{d+1}-n_1} \cdots x_{n_d}^{n_{d+1}-n_d} x_{d+1}^{n_{d+1}}.$$

The diagonal of this series is the subseries of terms where the exponents are the same. Thus $n_{d+1} = n_{d+1} - n_1$, for example, from which we deduce $n_{d+1} = n_1 = 0$. In fact, the only term with all exponents the same is the constant term, $f(0, \cdots, 0)$. □

We introduce a generalized extraction notation that frequently appears in combinatorial applications. For $F(\mathbf{x}) \in K[x_1, x_1^{-1}, \ldots, x_d, x_d^{-1}][[x_{d+1}]]$, define the **positive series extraction** as the sum of terms with non-negative coefficients. For example,

$$[x_1^{\geq 0} x_2^{\geq 0} \ldots x_d^{\geq 0}] F(\mathbf{x}) := \sum_{\mathbf{n} \in \mathbb{N}^d} f(\mathbf{n}) \mathbf{x}^{\mathbf{n}}.$$

We can express this subseries as a diagonal.

Proposition 3.2. *Suppose that* $F(\mathbf{x}) \in K[x_1, x_1^{-1}, \ldots, x_d, x_d^{-1}][[x_{d+1}]]$. *Then*

$$\left[[x_1^{\geq 0} x_2^{\geq 0} \ldots x_d^{\geq 0}] F(x_1, x_2, \ldots, x_{d+1}) \right]_{x_1 = \cdots = x_d = 1} = \Delta \frac{F\left(\frac{1}{x_1}, \frac{1}{x_2}, \ldots, \frac{1}{x_d}, x_1 x_2 \ldots x_{d+1}\right)}{(1 - x_1)(1 - x_2) \ldots (1 - x_d)}.$$

The key to this proof (which is left as an exercise) is the simple and useful generating function relation:

$$\frac{1}{1 - x} \sum_{n \geq 0} f_n x^n = \sum_{n \geq 0} \left(\sum_{k=0}^{n} f_k \right) x^n.$$

3.1.3   Algebraic Functions Are Diagonals

Where do diagonals appear in combinatorics? Let us start with algebraic series. Recall a series $F(x) \in K[[x]]$ is algebraic over K if there is a polynomial $P \in K[x, y]$, such that $P(x, F(x)) = 0$.

Proposition 3.3. *Let K be a field and $F(x) \in K[[x]]$ a algebraic formal power series with no constant term. Furthermore, assume that $P(x,y) \in K[x,y]$ is a polynomial, such that $P(x, F(x)) = 0$ and that $\partial P / \partial y(0,0) \neq 0$. Then the rational function*

$$R(x,y) := y^2 \frac{\frac{\partial P}{\partial y}(xy,y)}{P(xy,y)}$$

has a Taylor series expansion in $K[[x,y]]$ and satisfies

$$\Delta R = F.$$

Example 3.3. The generating function $C(x)$ for binary trees where size is give by number of vertices satisfies

$$C(x) = x + xC(x)^2 \implies P(x, C(x)) = 0 \text{ for } P(x,y) = x - y + xy^2.$$

As $P_y(x,y) = -1 + 2xy$, we verify $P_y(0,0) = -1$ and Proposition 3.3 applies. The conclusion is that

$$C(y) = \Delta y^2 \frac{-1 + 2xy^2}{xy - y + xy^3} = \Delta \frac{y(1 - 2y^2 x)}{1 - x(1 + y^2)}$$
$$= \Delta(1 + x + \ldots)y + (-x + x^3 + \ldots)y^3 + (-x^2 - x^3 + 2x^5 + \ldots)y^5$$
$$+ (-x^3 - 2x^4 - 2x^5 + 5x^7 + 14x^8 + \ldots)y^7 + \ldots.$$

An observation: the series expansion of this rational has positive and negative coefficients, that is, it is not combinatorial. Does it have a combinatorial interpretation?   ◄

The conditions on P in the proposition above appear frequently in the study of algebraic series. It is a matter only of a few straightforward polynomial manipulations to handle cases that are excluded by the hypotheses.

Under some similar conditions, an algebraic series in d variables can be expressed as the diagonal of a $2d$-dimensional rational function. However, in higher dimensions it is more difficult to avoid the analogous hypothesis on P.

3.1.4   Excursion Generating Functions

Diagonals appear frequently in lattice path combinatorics in the case of derived classes, and also when the step set has reflective symmetry. We identify the position $\mathbf{n} \in \mathbb{Z}^d$ in the integer lattice with the monomial $\mathbf{x^n}$. In this way, we encode movement on the lattice through Laurent polynomial multiplication. The number of walks of length n with steps from $\mathscr{S}$ from a to b restricted to the region C shall be denoted

$$\mathrm{walk}_C^{\mathscr{S}}(a \xrightarrow{n} b).$$

Either $\mathscr{S}$ or C may be omitted if they are clear by context. A walk model with fixed stepset $\mathscr{S} \subset \mathbb{Z}^d$ has **inventory polynomial** given by the Laurent polynomial $S(\mathbf{x})$:

$$S(\mathbf{x}) := \sum_{s \in \mathscr{S}} \mathbf{x}^s.$$

The product $S(\mathbf{x})^k$ indicates the possible end position after k steps since each lattice move is encoded by vector addition, which is an inherited, additive parameter. We can extract the number of walks that end at a given point by examining the coefficients of the appropriate monomial. In two dimensions this is

$$\mathrm{walk}_{\mathbb{Z}^2}^{\mathscr{S}}((k,\ell) \xrightarrow{n} (r,s)) = [x^r y^s] x^k y^\ell \left(\sum_{(s_1,s_2) \in \mathscr{S}} x^{s_1} y^{s_2} \right)^n.$$

Example 3.4 (Excursions). Consider the two-dimensional simple step set $\mathscr{S} = \{(\pm1,0),(0,\pm1)\} = \{\rightarrow,\leftarrow,\uparrow,\downarrow\}$. The inventory polynomial is $S(x,y) = x + \frac{1}{x} + y + \frac{1}{y}$. The walks that start and end at the origin are the **simple excursions**. Let $e(n)$ be the number of excursions of length n. The counting sequence $(e(2n)))_n$ starts

$$1,4,36,400,4900,\ldots \qquad \text{(OEIS A002894)}.$$

We can express this as a constant term extraction:

$$e(n) = \mathrm{walk}_{\mathbb{Z}^2}((0,0) \xrightarrow{n} (0,0)) = [x^0 y^0]\left(x + \frac{1}{x} + y + \frac{1}{y}\right)^n.$$

We deduce the generating function:

$$E(z) := \sum_{n \geq 0} e(n)z^n = \sum_{n \geq 0} \left(\mathrm{CT}\left(x + \frac{1}{x} + y + \frac{1}{y}\right)^n\right) z^n$$

$$= \mathrm{CT}_{x,y} \frac{1}{1 - z\left(x + \frac{1}{x} + y + \frac{1}{y}\right)}$$

$$= \Delta \frac{1}{1 - zxy\left(x + \frac{1}{x} + y + \frac{1}{y}\right)},$$

with the last equality a result of Proposition 3.1. Direct combinatorial argument can be used to show that $e(2n) = \binom{2n}{n}^2$. The combinatorial class of excursions is a subclass of the S-regular class $\{\rightarrow, \leftarrow, \uparrow, \downarrow\}^*$ defined by a parameter value so we say it is a derived class. In Exercise 1.11 it is proved that as a language it is not algebraic. Using tools from Chapter 5 we can prove the transcendence of the generating function using analytic arguments

The argument to enumerate excursions for any finite step set $\mathscr{S} \subset \mathbb{Z}^d$ is the same:

$$\sum_{n \geq 0} \text{walk}_{\mathbb{Z}^d}^{\mathscr{S}}((0,0) \xrightarrow{n} (0,0)) x_{d+1}^n = \Delta \frac{1}{1 - x_1 \ldots x_{d+1} S\left(\frac{1}{x_1}, \ldots, \frac{1}{x_d}\right)}. \tag{3.2}$$

◀

Example 3.5 (Delannoy numbers). Define $d_{n,m}$ to be the number of paths from the southwest corner $(0,0)$ of a rectangular grid to the northeast corner (m,n), using only single steps north, northeast, or east. These are called the **Delannoy numbers** after amateur French mathematician Henri Delannoy. When $n = m$ they are called the central Delannoy numbers. The sequence $(d_{n,n})_{n \geq 0}$ starts:

$$1, 3, 13, 63, 321, 1683, 8989, 48639, 265729, \ldots \qquad \text{(OEIS A001850)}.$$

We see that

$$d_{n,m} = [x^n y^m] \frac{1}{1 - (x + y + xy)}.$$

The generating function for the **central Delannoy numbers** is the diagonal:

$$D(y) = \sum_{n \geq 0} d_{n,n} y^n = \Delta \frac{1}{1 - (x + y + xy)}. \tag{3.3}$$

We can show the following binomial sum expression for the coefficients:

$$D(n) = \sum_{k=0}^{n} \binom{n}{k} \binom{n+k}{k}.$$

◀

Example 3.6 (Balanced classes). Let $\mathscr{B}$ be the class of binary words over $\{\circ, \bullet\}$ with no $\circ\circ\circ$ substring. Let $\mathscr{B}_=$ be the **balanced subclass** defined as the words that have the same number of $\circ$ as $\bullet$. The counting sequence starts

$$1, 2, 6, 16, 45, 126, 357, 1016, 2907, 8350 \qquad \text{(OEIS A005717)}$$

We mark occurrences of ◦ and ● by x and y, respectively. The generating function for $\mathscr{B}_=$ (by half length) is given by a diagonal

$$B_=(y) = \Delta \frac{1 + x + x^2}{1 - y(1 + x + x^2)}.$$

We remark, for any S-regular class $\mathscr{C}$, we can mark each type of atom by a separate size variable. The multivariable generating function for the full class is a multivariable $\mathbb{N}$-rational function. The subclass of balanced objects $\mathscr{C}_=$, wherein each atom occurs the same number of times (or in some fixed linear relation), is a derived class, and its generating function is a diagonal of a rational function. Balanced classes are algebraic when there are two types of atoms roughly because of two facts from formal language theory: (1) the language of words over an alphabet of size two where both letters appear with the same number of occurrences is a context-free language. (See Exercise 1.14.); (2) The intersection of a context-free language and a regular language is context-free. In this case, the generating function is

$$B_=(x) = \frac{2x}{1 - 2x - 3x^2 + (1 - x)\sqrt{1 - 2x - 3x^2}}.$$

However, if the alphabet has at least three elements, the above argument for algebraicity no longer holds, and many such derived balanced classes will not be context-free. It takes some additional argument to decide if the generating function is transcendental or not. However, we can easily define the diagonal in this case. (See Example 3.12.)   ◄

3.2   The Reflection Principle

Lattice walks with a particular symmetry in their step set can very often be enumerated using a strategy called the **reflection principle**. The basic idea is to assign a (possibly negative) weight to each object in such a way that the walks that leave the region are paired in a way so as to contribute a total weight of 0 towards the generating function. We obtain the generating function for the desired class by a subseries extraction on the weighted sum. The objects extracted from the general set form a derived class. Often the class is transcendental.

 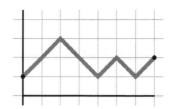

FIGURE 3.1
(left) A walk and its image under ϕ. *(right)* A fixed point of ϕ.

3.2.1 A One-dimensional Reflection

We illustrate the principle to find the generating function for Dyck paths. Consider the class W of walks, which begin at either $(0,1)$ or $(0,-1)$ and take steps from the set $\mathscr{S} = \{\nearrow, \searrow\} = \{(1,1),(1,-1)\}$. Note that this step set is symmetric across the x-axis. Mark the parameter end height with the variable u. Additionally we assign a weight based on the first step of the walk. The weight is $(+1)$ if the first step is $\nearrow$ and (-1) if it is $\searrow$.

The weighted generating function $W(u,x)$ for walks where length is marked by x and end height is marked by u:

$$W(u,x) = \sum_{n\geq 0}\sum_{k\in\mathbb{Z}} \text{walk}_{\mathbb{N}\times\mathbb{Z}}((1,1) \overset{n}{\longrightarrow} (n,k))\, u^k x^n \tag{3.4}$$

$$- \sum_{n\geq 0}\sum_{k\in\mathbb{Z}} \text{walk}_{\mathbb{N}\times\mathbb{Z}}((1,-1) \overset{n}{\longrightarrow} (n,k))\, u^k x^n \tag{3.5}$$

$$= \frac{u - \frac{1}{u}}{1 - x(u + \frac{1}{u})}. \tag{3.6}$$

We can re-interpret this expression. Consider the following sign-reversing involution ϕ: If a walk w never touches the axis, then $\phi(w) = w$. Otherwise ϕ flips the prefix of the walk up to the axis contact, and fixes the remainder of the walk. Because the step set is fixed under this action, the image of ϕ is indeed a valid walk. Figure 3.1 illustrates the two possibilities. Note, a walk and its (nontrivial) image will differ by a sign, but contribute the same (unsigned) monomial to the generating function. Since all such walks are in a unique pairing, the total contribution of the pair is 0. Therefore, the generating function only receives contributions from those walks which never touch the axis. There are two types: walks confined either above or below the axis.

The coefficient of u in $W(u,x)$ expanded first as a series in x counts walks with $\nearrow, \searrow$ steps that start and end at the same height and never go below that height. Those are precisely the Dyck paths. We can expand this explicitly to perform the extraction:

$$C(x) = [u]\frac{(u - \frac{1}{u})}{1 - x(u + \frac{1}{u})} \tag{3.7}$$

$$= [u]\frac{u}{1 - x(u + \frac{1}{u})} - [u]\frac{\frac{1}{u}}{1 - x(u + \frac{1}{u})} \tag{3.8}$$

$$= [u^0]\sum_{n} x^n\left(u + \frac{1}{u}\right)^n - [u^2]\sum_{n\geq 0} x^n\left(u + \frac{1}{u}\right)^n \tag{3.9}$$

$$= \sum_{n\geq 0} x^{2n}\left[\binom{2n}{n} - \binom{2n}{n+2}\right] \tag{3.10}$$

$$= \sum_{n\geq 0} \binom{2(n+1)}{n+1}\frac{1}{n+1}x^n. \tag{3.11}$$

We can also express this as a diagonal from Eq. (3.8):

$$C(x) = [u^0]\frac{1 - \frac{1}{u^2}}{1 - x(u + \frac{1}{u})} = \Delta\frac{1 - u^2}{1 - xu(u + \frac{1}{u})}.$$

Note that this is a very similar, but ultimately different diagonal expression for Catalan number than was given in Example 3.3: A given series can be the diagonal of many different rational functions. There are other analyses possible. We will see how to use Cauchy integrals in what is known as the diagonal method, and then we will see how determine the asymptotics of the coefficients.

Consider the set of walks with step set $\mathscr{S} = \{\nearrow, \searrow\}$ that stay above the x-axis but may end at any height. They are the set of prefixes of Dyck paths. We adapt the above argument to track other end heights.

The generating function $P(x)$ for this class is derived as follows:

$$P(x) := \sum_{n \geq 0} \sum_{k \geq 0} \text{walk}_{\mathbb{N}^2}((0,0) \xrightarrow{n} (n,k)) x^n \qquad (3.12)$$

$$= \sum_{k \geq 0} [u^{k+1}] \frac{(u - \frac{1}{u})}{1 - x(u + \frac{1}{u})} \qquad (3.13)$$

$$= \text{CT}_u \sum_{k \geq 0} \frac{1}{u^k} \frac{(u - \frac{1}{u})}{1 - x(u + \frac{1}{u})} \qquad (3.14)$$

$$= \text{CT}_u \frac{1/u}{1 - \frac{1}{u}} \frac{(u - \frac{1}{u})}{1 - x(u + \frac{1}{u})} \qquad (3.15)$$

$$= \Delta \frac{u}{1 - u} \frac{(u - \frac{1}{u})}{1 - xu(u + \frac{1}{u})}, \qquad (3.16)$$

with the last equality following from Proposition 3.1.

3.2.2   A Two-dimensional Reflection

This strategy generalizes to higher dimensions. We can determine a diagonal expression satisfied by the generating function of simple walks that start at the origin and remain inside the first quadrant $\mathbb{N}^2$. This shape of this region can tile the plane into four similar regions, bounded by the x and y axes. The simple step set is fixed by reflection across these axes, allowing us to adapt and apply the previous argument. We extract them from a class of weighted walks that start at one of four points and are unrestricted in the plane. Consider the set W of walks with step set $\mathscr{S} = \{(1,0), (-1,0), (0,1), (0,-1)\}$ that start at one of four points: $\{(\pm 1, \pm 1)\}$. Here the analysis follows the same structure once we define how to assign a sign, and give the sign reversing involution ϕ. The plane is divided into four quadrants, and the signs are assigned in an alternating fashion. The regions and signs are listed in Figure 3.2. If a walk w never touches an axis, we define $\phi(w) = w$. Otherwise, consider the first time w touches the axis. The image of w under ϕ is the walk that has this initial prefix mirrored across the touched axis, and remaining is the same. Note, the origin will never be the first touch to an axis so this is well-defined.

Simple excursions that remain in the first quadrant are equinumerous with simple walks that start and end at $(1,1)$ and never touch an axis. This can be directly seen in Figure 3.2, as it is a simple

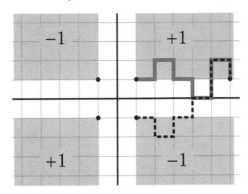

FIGURE 3.2
Four starting points for walks and how to pair up walks from neighbouring regions.

translation of the domain. We compute the generating function:

$$\sum_{n\geq 0}\mathrm{walk}_{\mathbb{N}^2}((0,0)\overset{n}{\longrightarrow}(0,0))z^n \tag{3.17}$$

$$= [x^1 y^1]\frac{xy - x/y + (xy)^{-1} + y/x}{(1 - z(x + 1/x + y + 1/y))} \tag{3.18}$$

$$= \mathrm{CT}\,\frac{\left(x - \frac{1}{x}\right)\left(y - \frac{1}{y}\right)}{xy(1 - z(x + 1/x + y + 1/y))} \tag{3.19}$$

$$= \Delta\frac{xy\left(x - \frac{1}{x}\right)\left(y - \frac{1}{y}\right)}{1 - zxy(x + 1/x + y + 1/y)} \tag{3.20}$$

$$= \Delta\frac{(x^2 - 1)(y^2 - 1)}{1 - z(x^2 y + y + xy^2 + x)}. \tag{3.21}$$

The generating function for walks that return to the x-axis is

$$\sum_{n\geq 0}\sum_{k\geq 0}\mathrm{walk}_{\mathbb{N}^2}((0,0)\overset{n}{\longrightarrow}(k,0))z^n = \Delta\frac{(x^2 - 1)(y^2 - 1)}{(1 - zxy(x + 1/x + y + 1/y))(1 - x)}.$$

The generating function for all walks in the quadrant is

$$\sum_{n\geq 0}\sum_{k\geq 0,\ell\geq 0}\text{walk}_{\mathbb{N}^2}((0,0)\xrightarrow{n}(k,\ell))z^n \tag{3.22}$$

$$=\Delta\frac{(x^2-1)(y^2-1)}{(1-zxy(x+1/x+y+1/y))(1-x)(1-y)} \tag{3.23}$$

$$=\Delta\frac{(x+1)(y+1)}{(1-zxy(x+1/x+y+1/y))}. \tag{3.24}$$

3.3 General Finite Reflection Groups

A strategy to enumerate multidimensional simple walks in the orthant $\mathbb{N}^d$ should be clear. Is that the only way this can generalize?

3.3.1 A Root Systems Primer

In fact, this is part of a general set-up of **finite reflection groups**. We will describe the framework and yield a different example. The key is that we divide $\mathbb{Z}^d$ into equal regions, and the step-set should have adequate symmetry as to support a reflection as we have done in the previous examples. The conditions that make this possible are very well-studied. A **root system** Φ is a set of roots, which are vectors in the Euclidean space that satisfies some particular geometrical properties. To ease the parsing of the definition, keep in mind the set $\{\pm e_1,\pm e_2\}$ and the integer lattice. Let E be a finite-dimensional Euclidean vector space, (such as $\mathbb{Z}^2$) with the standard Euclidean inner product denoted by $(\cdot,\cdot)$.

1. The roots span E;

2. The only scalar multiples of a root $\alpha\in\Phi$ that belong to ϕ are α itself and $-\alpha$;

3. For every root $\alpha\in\Phi$ the set Φ is closed under reflection through the hyperplane perpendicular to α;

4. If α and β are roots in Φ, then the projection of β onto the line through α is an integer or half-integer multiple of α.

A root system is minimally generated by a basis B.[1] The set of linear transformations generated by all the reflections with respect to the hyperplanes perpendicular to the roots is called the **Weyl group**, which we denote by $\mathcal{W}$. The root lattice $\mathcal{L}$ of a root system Φ is the $\mathbb{Z}$-submodule of E generated by Φ. It is a lattice in E. The **fundamental chamber** is the sublattice

$$\{x \in \mathcal{L} : (x, \alpha) > 0 \ \forall \alpha \in \Phi\}.$$

The length of an element σ of the Weyl group is denoted $\ell(\sigma)$ and is the least number of terms possible to express σ as a product of one of the fundamental reflections, σ_α, with $\alpha \in B$.

Example 3.7. If $\Phi = \{\pm e_1, \pm e_2\} \subset E = \mathbb{Z}^2$, then the Weyl group is

$$\begin{bmatrix} 1 & 0 \\ 0 & 1 \end{bmatrix}, \sigma_{e_1} = \begin{bmatrix} -1 & 0 \\ 0 & 1 \end{bmatrix}, \sigma_{e_2} = \begin{bmatrix} 1 & 0 \\ 0 & -1 \end{bmatrix}, \sigma_{e_1}\sigma_{e_2} = \begin{bmatrix} -1 & 0 \\ 0 & -1 \end{bmatrix}.$$

Their lengths are, respectively, 0, 1, 1 and 2. The fundamental region is the first quadrant.   ◀

We fix a finite set of steps, $\mathcal{S} \subset \mathcal{L}$. The number of walks of length n, from point a to b on this lattice with steps from $\mathcal{S}$ restricted to the fundamental chamber C is directly related to quantity of interest in representation theory. We shall denote this quantity as $\text{walk}_C(a \overset{n}{\longrightarrow} b)$. Let us see how to find it under some restrictions on $\mathcal{S}$.

A step set $\mathcal{S}$ is **reflectable** relative to a root system if it satisfies the following criteria:

1. $\mathcal{W} \cdot \mathcal{S} = \mathcal{S}$. That is, the Weyl group action applied to $\mathcal{S}$ fixes it;

2. For any α in Φ, the non-zero values of (α, s), as s ranges over $\mathcal{S}$, are $\pm k(\alpha)$, where $k(\alpha)$ is a fixed number that depends only on α. (This is a tricky technical condition to unpack, but it means that we cannot jump over boundaries.)

In our example root system, to be reflectable a step set must be symmetric with respect to both axes and be a subset of $\{\pm 1, 0\}^2$.

---

[1] Typically, the basis is denoted by the symbol Δ, but that would introduce too much confusion here.

3.3.2 Enumerating Reflectable Walks

In some cases, we can determine explicit generating functions for walks in fundamental chambers. Gessel and Zeilberger [GZ92] gave general formulas which hold under some symmetry conditions that assure that a reflection principle argument will work.

Theorem 3.4. *Assume that a and b are two lattice points that belong to the fundamental Weyl chamber. Then,*

$$\text{walk}_C^{\mathscr{S}}(a \xrightarrow{n} b) = \sum_{\sigma \in \mathscr{W}} (-1)^{l(\sigma)} \text{walk}_C(\sigma(a) \xrightarrow{n} b).$$

The proof uses precisely the reflection argument that we have seen so far. Note, we need a little bit of care for walks whose intersection is at an intersection of hyperplanes.

It is straightforward to see that in the unrestricted case:

$$\text{walk}_C(a \xrightarrow{n} b) = \text{CT} \frac{(\sum_s \mathbf{x}^s)^n}{\mathbf{x}^{b-a}}. \tag{3.25}$$

The number of walks of length n that go from a to b never touching the boundary of C is thus:

$$\text{walk}_C(a \xrightarrow{n} b) = \text{CT} \left[\frac{(\sum_{s \in \mathscr{S}} x^s)^n}{\mathbf{x}^{b-a}} \cdot \sum_{\sigma \in \mathscr{W}} (-1)^{\ell(\sigma)} \mathbf{x}^{\sigma(a)} \right]. \tag{3.26}$$

We can reformat this into a generating function:

$$W(x_{d+1}) = \text{CT} \left[\frac{1}{\mathbf{x}^{b-a}(1 - x_{d+1} \sum_{s \in \mathscr{S}} x^s)} \cdot \sum_{\sigma \in \mathscr{W}} (-1)^{\ell(\sigma)} \mathbf{x}^{\sigma(a)} \right].$$

For special values of a, the sum on the right side of this factors nicely. This is a consequence of the **Weyl denominator formula**. Let δ be one-half of the sum of all positive roots: $\delta = \frac{1}{2} \sum_{\alpha \in \Phi^+} \alpha$

$$\sum_{w \in W} (-1)^{\ell(w)} x^{w(\delta)} = x^{-\delta} \prod_{\alpha \in \Phi^+} (x^\alpha - 1).$$

We use this to rewrite the main formula.

Theorem 3.5 (Gessel and Zeilberger). *For any scalar c such that $c\delta$ is a lattice point, and for any lattice vector λ invariant under the Weyl group or every w in W, and such that $\lambda + c\delta, \alpha$ is an integral multiple of $k(\alpha)$ for every α in the basis of Φ,*

$$\text{walk}_C(\lambda + c\delta \xrightarrow{n} b) = \text{CT}_{x_1 \ldots x_d} \left[\frac{(\sum_s \mathbf{x}^s)^n}{\mathbf{x}^{b-a}} \cdot \mathbf{x}^{-\delta} \prod_{\alpha \in \Phi^+} (\mathbf{x}^{c\alpha} - 1) \right]. \quad (3.27)$$

Finally, the generating function for walks from $\lambda + c\delta$ to b remaining in the chamber is given by:

$$W(x_{d+1}) = \text{CT}_{x_1 \ldots x_d} \left[\frac{1}{\mathbf{x}^{b-a}(1 - x_{d+1} \sum_{s \in \mathscr{S}} \mathbf{x}^s)} \cdot \mathbf{x}^{-\delta} \prod_{\alpha \in \Phi^+} (\mathbf{x}^{c\alpha} - 1) \right].$$

3.3.3 A Non-simple Example: Walks in A_2

We can consider other Weyl groups to find richer families of lattice path models. The finite root system A_{n-1} consists of the $n(n-1)$ vectors

$$\Phi = \{e_i - e_j \mid 1 \le i \ne j \le n\}.$$

Its Weyl group is the symmetric group acting by permuting the coordinates, and its fundamental chamber is

$$C = \{(x_1, \ldots, x_n) \in \mathbb{R}^n \mid x_1 < \ldots < x_n\}.$$

A walk in the A_1 fundamental chamber is a one-dimensional simple walk. Next consider A_2. We remark that as the roots are linearly dependent, the model is two-dimensional. Relabel the root system as $\Phi = \{\pm\alpha_1, \pm\alpha_2, \pm(\alpha_1\alpha_2)\}$. Figure 3.3 illustrates how to choose vector α_1 so that the root system axioms hold. The projection of α_1 onto the line spanned by α_2 is $-1/2\alpha_2$ and vice versa. The basis B of the root system is α_1 and α_2. Label the reflection across the hyperplane perpendicular to α_i by σ_i. The dashed lines indicate the reflection hyperplanes; the shaded area is the fundamental chamber C.

It is convenient for our generating function purposes to apply a linear transformation in order to get integer exponents. The resulting image after transformation is given in Figure 3.4. The generating

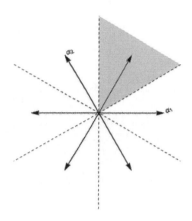

FIGURE 3.3
The root system A_2, and its fundamental chamber.

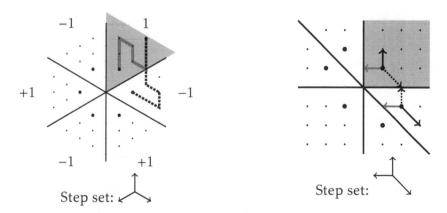

FIGURE 3.4
(left) Weyl chamber and lattice for A_2. A walk and its image under the involution ϕ. *(right)* The image of the regions in the rectangular lattice. A sample reflection of the step set into a neighbouring region.

function for walks that start and end at $(1,1)$ is

$$W(z) = \sum_{n \geq 0} \text{walk}_{\mathbb{N}^2}^{\{\uparrow,\searrow,\leftarrow\}}((0,0) \xrightarrow{n} (0,0)) z^n$$

$$= [x^1 y^1] \frac{xy - x^2/y + x/y^2 - 1/xy + y/x^2 - y^2/x}{1 - z(1/x + x/y + y)}$$

$$= \text{CT} \frac{xy - x^2/y + x/y^2 - 1/xy + y/x^2 - y^2/x}{xy(1 - t(1/x + x/y + y))}$$

$$= \Delta \frac{\left(1 - y^2/x + y^3 - x^2 y^2 + x^3 - x^2/y\right)}{(1 - z(y + x^2 + xy^2)))}.$$

The generating function for walks that end anywhere is:

$$\Delta \frac{\left(1 - y^2/x + y^3 - x^2 y^2 + x^3 - x^2/y\right)}{(1-x)(1-y)(1 - z(y + x^2 + xy^2)))}. \tag{3.28}$$

Now, it turns out that this generating function is algebraic. See Exercise 3.6.

3.4 Discussion

Diagonals appear throughout number theory, automata theory, physics and combinatorics [LvdP90, BBC+13, BR11].

The identification of diagonals of rational functions binomial sums is recent [BLS17].

For a proof of Proposition 3.3 see Denef and Lipshitz [DL87].

We have considered diagonals of series over fields of characteristic zero. There is an equally compelling life of diagonals in finite characteristics. See Adamczewski and Bell [AB12]. The behaviour is not the same: The diagonal of a multivariate algebraic power series with coefficients in a field of positive characteristic is algebraic [AB13].

Effective diagonals of algebraic functions were considered by Bostan Dumont and Salvy [BDS17]. Given a bivariate rational function, they consider algorithms for, and effective bounds on, algebraic equations satisfied by its diagonal.

Garrabrant and Pak describe an interpretation of diagonals of $\mathbb{N}$-rational series using an object called "irrational tilings" [GP]. They also show a correspondence between a family of binomial multisums (smaller than the binominal sums defined here) and diagonals of $\mathbb{N}$-rational functions. Is it possible to express the generating function of algebraic classes as diagonals of the multivariable generating function of an S-regular class equipped with an inherited multidimensional parameter? Garrabrant and Pak conjecture that cannot determine an $\mathbb{N}$-rational function whose diagonal grows asymptotically like the Catalan numbers [GP, Conjecture 11.2].

Root systems are a central topic in representation theory. They are classified into four infinite families and a collection of exceptional cases. The reader is pointed towards [Hum92].

Root systems and walks in their chamber have been enumerated since the early 1990s. The reflectable walks correspond to miniscule models. Grabiner's point of view is most similar to this. He finds exact expressions in terms of determinants of matrices with Bessel function entries. Kratthenthaler [Kra07] summarized many useful references in that text.

3.5   Problems

Exercise 3.1. Prove Proposition 3.2. ❐

Exercise 3.2. Show that the generating function for Motzkin paths is

$$\Delta\frac{x-x^{-1}}{1-y(x+1+x^{-1})}.$$

❐

Exercise 3.3 (Weighted walks). Assign a positive real number to each step set. The weight of a walk is the product of the weights of its steps. Determine an expression for the generating function of d-dimensional weighted simple excursions. Be explicit about any conditions on the weights you require. ❐

Exercise 3.4 (Diagonal steps). Perform the same procedure to find the generating functions for diagonal walks. These have step set $\{\pm e_1 \pm e_2\}$. Be careful with your treatment of the origin! ❐

Exercise 3.5 (Simple walks in arbitrary dimension). It is straightforward to express a higher dimensional analogue: Find the generating function for walks starting and ending at the origin, restricted to the first orthant with steps from $\{\pm e_i : 1 \leq i \leq d\}$. This appeared in the literature in the 1990s. The numerator is compactly expressed using a determinant. Adding weights to the directions is also a straightforward affair. It becomes less clear what to do with the expressions however. Determine an expression for the generating function of d-dimensional weighted simple excursions.

Show that the generating function for the d-dimensional simple walks are given by

$$\Delta \frac{G(x_1,\ldots,x_d,t)}{H(x_1,\ldots,x_d,t)} = \frac{(1+x_1)\cdots(1+x_d)}{1 - t\sum_{k=1}^{d}(1+x_k^2)(z_1\cdots x_{k-1}x_{k+1}\cdots x_d)}.$$

Reference: [Zei83] [Zei90] and [MM16].

❏

Exercise 3.6 (Standard Young tableau of bounded height). Stanley, in his foundational combinatorial study of D-finite functions, asked if the class of standard Young tableau of bounded height had a D-finite generating function. This is a class of objects defined on Young diagrams.

1. Prove that the A_2 walks are in bijection with standard Young tableau of bounded height.

2. Prove that standard Young tableaux of bounded height are D-finite for a fixed height. Hint, build part 1 into a bijection to lattice walks.

Indeed, there are a surprising number of different bijections between standard Young tableaux of bounded height and lattice walks. Several are useful for deducing D-finiteness.

Reference: [Mis19].

❏

Exercise 3.7. Prove that the class of involutions with no increasing subsequence of length k is D-finite. ❏

Exercise 3.8. Prove that the set of $0-1$ matrices such that each row and column sum is equal to k is D-finite. ❏

Exercise 3.9. Deduce from the extraction the following formula for simple excursions restricted to the first quadrant:

$$e(2n) = \sum_{k=0}^{n} \binom{2n}{k} c_k c_{2n-k},$$ (3.29)

where c_k is the k-th Catalan number. ❏

Exercise 3.10. We can define higher dimensional central Delannoy numbers:

$$\sum_{\mathbf{n} \in \mathbb{N}^d} D(\mathbf{n}) x_d^n = \Delta \frac{1}{2 - \prod_{k=1}^{d} (1 + x_k)}.$$ (3.30)

Determine an expression for $D(\mathbf{n})$ in dimension 3, 4, ...

Reference: [CDNS11].

❏

Exercise 3.11. List all 2D reflectable step sets for the root system defined by $\Phi = \{\pm e_1, \pm e_2\}$. How many are there of dimension d?

Reference: [MM16].

❏

Exercise 3.12. Let $\mathscr{T}$ be the class of words over $\{a,b,c\}$ with no *aaa*, nor *cbc* substring. Determine a diagonal expression for the generating function of balanced words in this class. (That is, the subclass of words such that $\#a's = \#bs = \#cs$.) Develop your series to verify your result. ❏

Exercise 3.13. Determine a grammar for non-empty Motzkin paths. Use this to express the generating function as a diagonal of a rational function. ❏

Part II

Methods for Asymptotic Analysis

4

Generating Functions as Analytic Objects

CONTENTS

4.1	Series Expansions	82
	4.1.1 Convergence	82
	4.1.2 Singularities	83
4.2	Poles and Laurent Expansions	84
	4.2.1 Puiseux Expansions	86
4.3	The Exponential Growth of Coefficients	87
4.4	Finding Singularities	91
4.5	Complex Analysis	92
	4.5.1 Primer on Contour Integrals	92
	4.5.2 The Residue of a Function at a Point	93
4.6	Asymptotic Estimates for Meromorphic Functions	95
4.7	The Transfer Lemma	98
4.8	A General Process for Coefficient Analysis	99
4.9	Multiple Dominant Singularities	102
4.10	Saddle Point Estimation	106
4.11	Discussion	108
4.12	Problems	108

We can deduce very precise asymptotic estimates about counting sequences by viewing formal series as analytic functions. The growth of the coefficients of a series is directly related to how the function behaves at its singularities. To understand the connection, we first discuss convergence of formal power series, and then consider the implications of writing coefficients as complex contour integrals. Estimating contour integrals is the centerpiece of analytic combinatorics. The fundamentals that we review here form the basis of the multivariable case. Thus, even if you are familiar with these results, it is useful to have them in mind as a foundation for the more general setting.

4.1   Series Expansions

Power series are not only formal objects that constitute a ring: They can be evaluated at points to define functions. To emphasize that we work on complex analytic functions, we write our series as functions of z in this chapter.

4.1.1   Convergence

A series $\sum_{n\geq 0} f_n z^n$ is said to be **convergent at a point** z_0 if the sequence of partial sums

$$\left(\sum_{n=0}^{N} f_n z_0^n \right)_N$$

converges to a finite limit. A convergent series is further qualified to be **absolutely convergent** if the series $\sum_{n=0}^{\infty} f_n |z|^n$ is also convergent.

The geometric series $\sum_{n\geq 0} z^n$ is convergent when $|z| < 1$ and converges to the value $\frac{1}{1-z}$. This series is absolutely convergent inside this disc. The sequence of partial sums tends to infinity at $z = 1$, and at $z = -1$ the sequence of partial sums oscillates between 1 and 0, hence it fails to converge at $z = -1$. The exponential series $\sum z^n/n!$ is convergent for every complex $z \in \mathbb{C}$ and takes the value e^z.

The **Weierstrass M-test** is a useful tool to understand if a series converges absolutely.

Proposition 4.1 (Weierstrass M-test). *Suppose that $(f_n(z))_n$ is a sequence of real or complex valued functions on set Ω, and that there is a sequence of positive numbers (M_n) satisfying $|f_n(z)| \leq M_n$, for all z in Ω, and additionally, $\sum_{n\geq 0} M_n$ converges to a finite value. Then, the series $\sum_{n\geq 0} f_n(z)$ converges absolutely and uniformly on Ω.*

From this proposition we conclude that if the power series $\sum_{n\geq 0} f_n z^n$ is convergent at $z = r$, for positive real r, it is absolutely convergent for all $z \in \mathbb{C}$ satisfying $|z| \leq r$. The **radius of convergence of a series** is the radius of the largest (open) disc in which the series converges. We shall denote the radius of convergence of series $F(z) = \sum_{n\geq 0} f_n z^n$ is sometimes denoted $\mathrm{ROC}(F(z))$ or R. A power series is absolutely convergent for all points in the **open disc of convergence**, $\{z \mid |z| < R\}$. The set

$$\{z \mid |z| = R\}$$

is the **circle of convergence**.

We also consider series developments centred at other points. The power series $\sum_{n \geq 0} f_n (z - z_0)^n$ is said to be centred at z_0. It is absolutely convergent in some disc

$$\{z \mid |z - z_0| = R\}.$$

4.1.2 Singularities

A series $F(z) = \sum_{n \geq 0} f_n (z - z_0)^n$ is said to be **analytic** at z_0 if its radius of convergence is positive. That is, it is convergent in some nontrivial open disc around z_0.

Analytic functions have a second characterization in terms of differentiation. Complex differentiation is defined analogously to the real valued case:

$$F'(z) := \lim_{h \to 0} \left| \frac{F(z + h) - F(z)}{h} \right|. \tag{4.1}$$

If this limit exists at $z = z_0$, we say the function has a derivative at z_0. If F has a derivative for every point in an open neighbourhood of z_0 then it is **holomorphic** at z_0. If a function is holomorphic for all of $\mathbb{C}$ then it is **entire**. The exponential function is an example of an entire function.

A classic result of complex analysis identifies holomorphic and analytic functions: A function is holomorphic at z_0 if and only if it is analytic at z_0.

A function is **singular at a point** z_0 if it is not analytic at z_0 but is analytic in a neighbourhood of z_0. A singularity z_0 of $F(z)$ is **isolated** if $F(z)$ is analytic in an open set around z_0 with z_0 removed, for example,

$$\{z \mid 0 < |z - z_0| < R\}.$$

This is a **punctured neighbourhood of** z_0.

Example 4.1. Consider the function $F(z) = \frac{1}{1-z}$. This is not defined at $z = 1$, hence it cannot be differentiable at that point; there is no series expansion $F(z) = \sum_{n \geq 0} f_n (z - 1)^n$ with a non-zero radius of convergence. A derivative does exist at $z = -1$, and we can find a Taylor expansion around $z = -1$:

$$\frac{1}{1-z} = \sum_{n \geq 0} \frac{1}{2^{n+1}} (z + 1)^n \quad \text{when } |z + 1| < 1.$$

Thus, $F(z)$ is analytic at $z = -1$. Now, the geometric series $\sum_{n\geq 0} z^n$ is not convergent at -1. When we talk about the analytic points of a function, we must distinguish this from the convergence of a particular series representation of that function.

We note also, that there is a series expression centered at 0 that is valid beyond $|z| = 1$:

$$\frac{1}{1-z} = \frac{1}{z(\frac{1}{z}-1)} = \frac{-1/z}{1-\frac{1}{z}} = \sum_{n\geq 0} \frac{-1}{z}^{\,n+1} \quad \text{when } |z| > 1.$$

This is not the Taylor expansion but is a valid series expansion. It also helps emphasize the point that we must be clear when we write $[z^n]F(z)$, precisely the series to which we refer. ◄

A function that is analytic in some open disc may have singularities so densely packed on the boundary of convergence that for any point on the boundary, all neighbourhoods contain a singularity. This is an example of a **natural boundary**. The existence of a natural boundary can indicate a complicated complex analytic structure.

Example 4.2. The function $F(z) = \prod_{k\geq 0} \frac{1}{1-z^k}$ is singular at all rational roots of unity, $e^{r\pi i}, r \in \mathbb{Q}$. These are dense on the unit circle, and so the function has a natural boundary at the unit circle. ◄

4.2 Poles and Laurent Expansions

We have seen that a function $F(z)$ is analytic at $z_0 \in \Omega$ when $F(z)$ has a power series representation

$$F(z) = \sum_{n\geq 0} c_n (z - z_0)^n,$$

which is valid for z in some open disc about z_0 contained in Ω. In combinatorics, we mostly consider Taylor series but we also frequently develop functions around other points, especially their singularities. Is there a series representation when $F(z)$ is not analytic at z_0? The answer depends on the nature of the singularity at z_0.

A **pole** of a complex function is a singularity that behaves like $1/z$ at $z = 0$. A function $F(z)$ has a pole of order M at z_0 if one can write

$$F(z) = \frac{G(z)}{(z - z_0)^M},\tag{4.2}$$

where $G(z)$ is analytic and non-zero at z_0, and M is a positive integer. Thus, the value of $F(z)$ diverges towards infinity as one approaches z_0. Rational functions have poles at the roots of their denominator.

A function is said to be **meromorphic** if it only has poles for singularities. A plot of $|R(z)|$ illustrates a pole z_0. Then, $F(z)$ admits a series expansion using negative powers of $(z - z_0)$. It is known as a **Laurent expansion** and looks like

$$F(z) = \frac{f_{-M}}{(z - z_0)^M} + \cdots + f_0 + f_1(z - z_0) + f_1(z - z_0)^2 + \cdots.$$

If $f_{-M} \neq 0$ and $M \geq 1$ then we say $F(z)$ has a **pole of order** M at z_0. It is an example of a Laurent series. This is reasonable: It suffices to replace $G(z)$ in Eq. (4.2) by its Taylor series at z_0 and divide through by $(z - z_0)^M$.

The domain of convergence of a Laurent expansion is an **annulus**. An annulus is an open region defined by an inner and outer radius

$$\{z \in \mathbb{C} \mid r < |z - z_0| < R\}.$$

It can be a punctured neighbourhood ($r = 0$), the punctured plane ($R = \infty$) or simply a ring shape. A function can have series expansions centered at z_0, each unique for a fixed annulus of convergence. In a region without singularities, it is precisely the Taylor expansion.

Example 4.3. Consider the function

$$R(z) = \frac{1}{(1 - z^3)(1 - 4z)^2(1 - 5z^4)}.$$

It has a Taylor series expansion around the origin valid up to $1/4$. Since for points near $z = 1/4$ we can write

$$R(z) = \frac{(1 - z^3)^{-1}(1 - 5z^4)^{-1}}{(1 - 4z)^2} = \frac{1024}{15813} \frac{1}{(z - 1/4)^2} + O((z - 1/4)^{-1})$$

we conclude that $R(z)$ as a pole of order 2 at $z = 1/4$. The other seven poles are all of order 1.

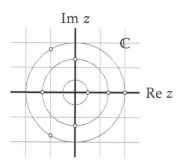 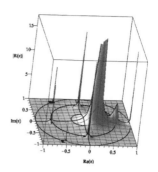

FIGURE 4.1

(left) The eight complex singular points of $\frac{1}{(1-z^3)(1-4z)^2(1-5z^4)}$ from Example 4.3. *(right)* The divergence at the poles.

Figure 4.1 (left) is a plot of the singularities of $R(z)$, each delimiting an annulus where there would be a unique series expansion centered at 0. On the right we see the numerical evidence that these points are poles. As we see in the right of the figure, the value of $|R(z)|$ diverges as these points – the large spike (shaped like a pole) is behaviour of a function at a pole. We can see that the pole of order 2 has a stronger divergence.   ◁

4.2.1 Puiseux Expansions

Another type of singularity common in combinatorial analysis is a **branch point**. A function fails to be analytic at a branch point not because of an issue of divergence, but because there is no series expansion valid in an open disc around the point – one must always delete a ray from the branch point. The problem stems from how logarithms, and consequently non-integral powers, are defined. For example, $z^{1/2}$ has a branch point at $z = 0$. Branch points arise when we consider solutions to algebraic equations.

In this case, we use may use Puiseux expansions. **Puiseux series** use not only negative but also fractional exponents.

In particular, if a function $F(z)$ is algebraic, and satisfies a polynomial equation $P(z, F(z)) = 0$, then $F(z)$ has a Puiseux expansion that is convergent in a neighbourhood of 0. Indeed, all of the solutions to the polynomial equation will have a Puiseux expansion, and understanding how they relate is important to the study of algebraic curves. Here, we will not explain how to determine such series expansions, as typically they are quite accessible via computer algebra programs. For combinatorial equations, they can be obtained by iterating the functional equation.

Example 4.4 (Ternary trees). Consider the family of **ternary trees** defined by the combinatorial specification

$$\mathcal{T} \equiv \bullet \times \left(\epsilon + \mathcal{T}^3 \right) \implies T(z) = z \cdot (1 + T(z)^3).$$

The polynomial equation $y - z(1 + y^3) = 0$ has three series solutions[1]:

$$y = z + z^4 + 3z^7 + O(z^8) \tag{4.3}$$

$$y = \frac{1}{\sqrt{z}} - z/2 - 3/8\, z^{5/2} + O\left(z^3\right) \tag{4.4}$$

$$y = -\frac{1}{\sqrt{z}} - z/2 + 3/8\, z^{5/2} + O\left(z^3\right). \tag{4.5}$$

Clearly, the first solution is the generating function answer we are looking for. It is the solution we obtain if we iterate the system $y_k = z(1 + y_{k-1}^3)$ with initial condition $y_0 = 0$. The other two are examples of Puiseaux series expansions. ◄

4.3 The Exponential Growth of Coefficients

A singularity is a point at which a function ceases to be analytic. It could be a pole; it could be a branch point. As power series is convergent inside its disc of convergence, it is is analytic there. Thus, a **series cannot have any singularity inside its disc of convergence.** If a series is not entire, we know that it must have at least one singularity on its radius of convergence.

Proposition 4.2. *A function $F(z)$, analytic at the origin, whose series expansion about 0 has a finite radius of convergence R must have a singularity on $|z| = R$.*

Singularities on the circle of convergence are called **dominant singularities**. If there is a unique dominant singularity, it is a **minimal**

---

[1] These were obtained with the `algeqtoseries` procedure from the Maple package gfun[SZ94].

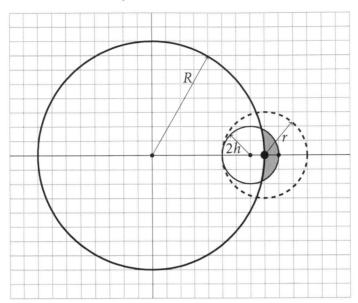

FIGURE 4.2

The different domains in the proof of Pringsheim's lemma.

dominant singularity. We can say something more precise in the case of combinatorial generating functions, as they have the additional characteristic that all the coefficients are positive.

Lemma 4.3 (Pringsheim's lemma). *Suppose that $F(z)$ is analytic with series expansion*

$$F(z) = \sum_{n=1}^{\infty} f_n z^n \quad with \quad f_n \geq 0.$$

If the series has radius of convergence R, then the point $z = R$ is a singularity of $F(z)$.

The proof is quite insightful – we see how the positivity of the coefficients is important in the localization of the dominant singularity.

Proof. Let us suppose, towards a contradiction, that F is actually analytic at $z = R$. Then, by definition, it must also be analytic in some small disc about $|z - R| < r$. Consider $0 < h < r/3$ and expand F around $z_0 = R - h$:

$$F(z) = \sum_{k \geq 0} g_k (z - z_0)^k.$$

We can find an explicit expression for g_k in terms of the f_n using the formula for Taylor expansion:

$$g_k = \sum_{n \geq 0} \binom{n}{k} f_n z_0^{n-k}.$$

Since the f_n are positive, we deduce that all of the g_k are nonnegative. It also gives us that the following expansion valid is some region around z_0:

$$F(z) = \sum_{k \geq 0} (z - z_0)^k \sum_{n \geq 0} \binom{n}{k} f_n z_0^{n-k}.$$

Now since F is analytic in the disc $|z - R| < r$, this expansion for F about z_0 must be convergent at $z = R + h$. We can evaluate it at this point:

$$F(R + h) = \sum_{k \geq 0} (2h)^k \sum_{n \geq 0} \binom{n}{k} f_n z_0^{n-k}.$$

This is a key point: As this is a convergent double sum of positive terms, the terms may be reordered:

$$F(R + h) = \sum_{n \geq 0} f_n \sum_{k \geq 0} \binom{n}{k} z_0^{n-k} (2h)^k$$

$$= \sum_{n \geq 0} f_n (z_0 + 2h)^n$$

$$= \sum_{n \geq 0} f_n (R + h)^n.$$

If this sum to be convergent, then $F(z)$ is convergent at $R + h$. In Figure 4.2 we deduce points in the grey region for which the series at the origin is also convergent, but this contradicts the assumption that the radius of convergence is exactly R. There is no point in this region for which the series is convergent. Thus, $F(z)$ is not analytic at $z = R$. □

The dominant singularity of a combinatorial generating function is along the positive real axis. Furthermore, this singularity dictates the exponential growth of the coefficients.

A sequence $(a_n)_n$ is said to be of **exponential order** K^n if and only if

$$\limsup_{n \to \inf} |a_n|^{1/n} = K.$$

In this case, we have for any $\epsilon > 0$,

- $|a_n| > (K - \epsilon)^n$ for infinitely many values of n, and

- $|a_n| < (K + \epsilon)^n$ for all but a finite number of values of n.

Theorem 4.4. *If $F(z)$ is analytic at 0, and R is the modulus of the singularity closest to the origin (so that $R = \sup \{r \geq 0 : F(z)$ is analytic in $|z| < r\})$ then we have*

$$\limsup_{n \to \infty} f_n^{1/n} = R^{-1}.$$

In the case that the coefficients are nonnegative, then we have

$$R = \sup\{r \geq 0 : F(z) \text{ is analytic at all } 0 \leq z < r\}.$$

Proof. Let R be the radius of convergence of F about the origin. Since F is convergent everywhere inside $|z| < \rho$, we have $R \geq \rho$. And since there must be a singularity somewhere on $|z| = \rho$, we must have $R = \rho$. Now let $\epsilon > 0$. By definition of the radius of convergence, we have $f_n(R-\epsilon)^n \to 0$, and hence $|f_n|^{1/n} < (R-\epsilon)^{-1}$ almost everywhere. If there is some M such that $|f_n|(R+\epsilon)^n < M$ for all n, then

$$|f_n| (R + \epsilon/2)^n < M \left(\frac{R + \epsilon/2}{R + \epsilon} \right)^n \to 0.$$

Furthermore,

$$\sum_{n \geq 0} |f_n| (R + \epsilon/2)^n < \sum_{n \geq 0} M \left(\frac{R + \epsilon/2}{R + \epsilon} \right)^n$$

which is convergent by the Weierstrass M-test since $R + \epsilon/2 < R + \epsilon$. **This contradicts our assumption about the radius of convergence!** We conclude that $|f_n|^{1/n} > (R + \epsilon)^{-1}$ infinitely often. Thus $\limsup_{n \to \infty} f_n = R^{-1}$. $\square$

When a function has multiple singularities, one is typically only interested in the dominant ones as the effect on the asymptotics of the others decays exponentially.

This gives us exactly the first principle of coefficient asymptotics.

First Principle of Coefficient Asymptotics

The **location** of singularities of an analytic function determines the **exponential order** of growth of its Taylor coefficients.

Example 4.5 (Motzkin paths). Recall that the generating function of Motzkin paths is $M(z) = \sum_{n\geq 0} m_n z^n = \frac{1-z-\sqrt{1-2z-3z^2}}{2z}$. Let us look for singularities. Initially, we guess there is a pole at $z = 0$ and two branch points, the solutions to $1 - 2z - 3z^2 = 0$. Now, the numerator is 0 at $z = 0$, hence its Taylor series has no constant term; $M(z)$ is analytic at 0. This leaves the branch points. We solve $1 - 2z - 3z^2 = 0 = (1 + z)(1-3z)$. The dominant singularity is at $z = 1/3$. Thus we conclude the Motzkin numbers satisfy $\limsup_{n\to\infty} m_n^{1/n} = 3$. This seems reasonable, since there are at most three possibilities for each step. ◄

4.4   Finding Singularities

Given an explicit function we now know to look for poles and branch points. However, we can also find singularities without solving for explicit expressions for the generating function. There are some combinatorial arguments for finding the dominant singularities that might be applicable to combinatorial generating functions. For a combinatorial class $\mathscr{C}$, let ρ_C be the real valued dominant singularity (radius of convergence) of its OGF. Table 4.1 summarizes the impact on the singularities

We finish the table by considering $C(z) = \frac{1}{1-A(z)}$ a little closer. For this to be well-defined, we must have $A(0) = 0$. If $\lim_{z\to\rho_A} A(z)$ is infinite, then since the function is increasing, and the dominant singularity is on the positive real axis, there is some z_0 such that $A(z_0) = 1$. Then we have that $\rho_C = z_0$, and $\lim_{z\to\rho_C} C(z)$ is infinite. We note additionally that z_0 is always computable to arbitrary precision using, say, binary search.

Proposition 4.5 (Regularity criterion). *The generating function of a non-trivial S-regular class is divergent at its dominant singularity.*

TABLE 4.1
Admissible Operators and Dominant Singularities

$\mathscr{C}$	$C(z)$	ρ_C
$\mathscr{E}$	1	∞
$\mathscr{Z}$	z	∞
$\mathscr{A} + \mathscr{B}$	$A + B$	$\min\{\rho_A, \rho_B\}$
$\mathscr{A} \times \mathscr{B}$	$A(z)B(z)$	$\min\{\rho_A, \rho_B\}$
$\mathscr{A}^*$	$\frac{1}{1-A(z)}$	ρ_C satisfies $A(\rho_C) = 1$

A corollary of this is another proof that Motzkin paths and binary trees can not be specified by an S-regular specification. Both of these classes have generating functions that are finite when evaluated at their dominant singularities.

4.5   Complex Analysis

Classic complex analysis of contour integrals permits a second approach to study coefficients of series expansions. Given a function, we are able to write the coefficient $[z^n]F(z)$ of a Laurent series expansion of $F(z)$ as an integral. We will have fast methods to evaluate and estimate the integrals.

4.5.1   Primer on Contour Integrals

A **contour** γ in $\mathbb{C}$ is a smooth curve. A contour is **simple** if it never intersects itself. A contour may be parameterized $\gamma \equiv \gamma(t)$ for $t \in [0, 1]$. If $\gamma(0) = \gamma(1)$, we say that the contour is closed. A **contour integral** is defined for a complex function, and a contour integral in terms of a function on a real domain:

$$\int_\gamma F(z)\,dz := \int_0^1 F(\gamma(t))\gamma'(t)\,dt.$$

An important feature is that there are many results that give the value of a contour integral without passing through classic methods of integral calculus. We summarize the key points here, but most are not too difficult to establish from the definition. The value of the integral is independent of the parametrization. A most remarkable, and useful feature of contour integrals is that the contour can be continuously deformed, and under some conditions, the value of the integral does not change. Judicious manipulation of the contour is a central strategy of analytic combinatorics.

Theorem 4.6. *Let $F(z)$ be analytic in $\Omega \subset \mathbb{C}$, and let γ be a simple loop in Ω. Then*

$$\int_\gamma F(z)\,dz = 0.$$

If $F(z)$ has a continuous anti-derivative, then $\int_\gamma F(z)dz = 0$ for any closed contour.

Example 4.6. Let γ be the contour that is the small circle around the origin parametrized as $\gamma(t) = e^{2\pi i t}, t \in [0,1]$. Then

$$\int_\gamma \frac{1}{z}\,dz = \int_0^1 e^{-2\pi i t} \cdot 2\pi i \cdot e^{2\pi i t}\,dt = 2\pi i \int_0^1 1\,dt = 2\pi i.$$

As z^k is analytic for positive k, $\int_\gamma z^k\,dz = 0$. For negative powers, aside from when $k = -1$, $\int_\gamma z^k\,dz = 0$ since the integrand has a continuous anti-derivative. Note that γ can be a circle of any non-zero radius, and these values are the same. ◀

4.5.2 The Residue of a Function at a Point

The **residue of f at z_0** is defined as the coefficient of $(z - z_0)^{-1}$ in the Laurent series expansion around z_0. That is, if

$$F(z) = \sum_{n=-M}^{\infty} f_n(z - z_0)^n,$$

then the residue of $F(z)$ at z_0 is f_{-1}. The residue is denoted $\operatorname{Res}_{z=z_0} F(z)$. The residue of an entire function is zero, as at any point the function is analytic, and hence has a Taylor expansion at that point, which has only terms with positive powers.

Example 4.7. To compute $\mathrm{Res}_{z=1} \frac{z}{1-z}$, we first do the series expansion of the numerator around 1, and then simplify. $\frac{z}{1-z} = -\frac{z-1+1}{z-1} = -\frac{z-1}{z-1} - \frac{1}{z-1}$. We compute $\mathrm{Res}_1 \frac{z}{1-z} = -1$. ◂

Many formulas for computing residues of meromorphic functions can be deduced by simple series manipulations. For example, suppose F has a simple pole at z_0, and G is analytic at z_0. Then $\mathrm{res}_{z_0}(FG) = G(z_0)\,\mathrm{res}_{z_0}(F)$.

The following result (up to some details about convergence, and a small change of variables) is a result of Example 4.6. Roughly, only the residue term contributes to the integral.

Proposition 4.7 (Local residue formula). *Let z_0 be an isolated singularity of $F(z)$ and let γ be a small circle centered at z_0, such that F is **analytic** on γ and its interior, except possibly at z_0. Then,*

$$\int_\gamma F(z)\,dz = 2\pi i\,\mathrm{Res}_{z=z_0} F(z).$$

From this we can apply Cauchy's residue theorem.

Theorem 4.8 (Cauchy's residue theorem). *Let $F(z)$ be meromorphic in Ω, and let γ be a positively oriented simple loop in Ω along which $F(z)$ is analytic. Then*

$$\frac{1}{2\pi i}\int_\gamma F(z)\,dz = \sum_s \mathrm{Res}_{z=s} F(z),$$

where the sum is over all poles s of F inside γ.

We use the residue formula in the opposite sense: It is a means to extract any coefficient from a series expansion.

Theorem 4.9 (Cauchy's integral formula). *Let $F(z)$ be analytic in a region Ω that contains 0, and let γ be a positively-oriented simple loop in Ω that encloses 0. Then*

$$[z^n]F(z) = \frac{1}{2\pi i}\int_\gamma F(z)\,\frac{dz}{z^{n+1}}.$$

Cauchy's Integral Formula (CIF)

$$[z^n]F(z) = \frac{1}{2\pi i} \int_\gamma F(z) \frac{dz}{z^{n+1}}.$$

4.6 Asymptotic Estimates for Meromorphic Functions

We can access the coefficients of rational function using only three tools: partial fraction decomposition, geometric series and generalized binomials. This gives a complete picture of the form of the coefficients.

Theorem 4.10 (Rational function asymptotics). *Let $F(z)$ be a rational function that is analytic at zero with poles at $\alpha_1, \ldots, \alpha_m$. Then there exist polynomials $p_j(n), j = 1, \ldots, m$ such that for n larger than some n_0*

$$[z^n]F(z) = \sum_{j=1}^m p_j(n)\alpha_j^{-n},$$

where the degree of p_j is the order of the pole at α_j minus one.

In order to compute the asymptotic growth of the coefficients, it is sufficient to consider the poles of smallest modulus, and the subexponential growth is given by the order of that pole. The error term is given by the second closest pole.

Meromorphic functions have a similar behaviour near their singularities, and we can deduce a similar formula. It is straightforward to build a strong estimate using Cauchy integrals and residue computations.

Theorem 4.11 (Meromorphic function asymptotics). *Let $F(z)$ be a function that is meromorphic on the closed disc $|z| \le R$ with poles at $\alpha_1, \ldots, \alpha_m$. Assume that F is analytic on $|z| = R$ and at $z = 0$. Then there*

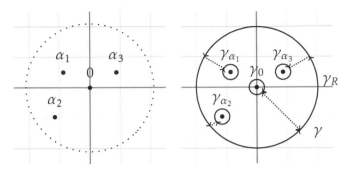

FIGURE 4.3
Possible poles of $F(z)/z^{n+1}$ and the contour γ to build for the proof of
Theorem 4.11.

exist polynomials $p_j(n), j = 1,\ldots, m,$ *such that*

$$[z^n]F(z) = \sum_{j=1}^{m} p_j(n)\alpha_j^{-n} + O(R^{-n}),$$

where the degree of p_j *is the order of the pole at* α_j *minus one.*

Proof. We define γ to be the curve with several components, visualized in Figure 4.3.

1. γ_R: The counter clockwise circle of radius R

2. γ_s: For s either a pole of F or 0, γ_s is the clockwise circle around s
   which contains no other poles

3. Line integrals from γ_R to γ_s, and then the reverse back.

We can break γ down into its components and build an equation by
computing the contribution from each contour segment. Note that the
interior of γ contains no singularities and hence

$$\int_{\gamma} F(z)/z^{n+1}\,dz = 0.$$

By a straightforward estimate

$$\int_{\gamma_R} F(z)/z^{n+1}\,dz = O(R^{-n}).$$

Since the line integrals occur in cancelling pairs, they contribute 0. Thus,

$$0 = \int_\gamma F(z)/z^{n+1}\,dz = \int_{\gamma_R} F(z)/z^{n+1}\,dz + \sum_{s\in\{0,\alpha_1,\dots,\alpha_m\}} \int_{\gamma_s} F(z)/z^{n+1}\,dz$$

$$= O(R^{-n}) + \underbrace{\mathrm{Res}_{z=0}\frac{F(z)}{z^{n+1}} + \sum_{s\in\{\alpha_1,\dots,\alpha_m\}} \mathrm{Res}_{z=s}\frac{F(z)}{z^{n+1}}}_{f_n}.$$

The residue at a pole α is computed using the Taylor expansion of z^{-n-1} at α. Say α is a pole of order M,

$$F(z) = \frac{c}{(z-\alpha)^M} + \dots$$

Then,

$$\mathrm{Res}_{z=\alpha} F(z)\, z^{-n-1} = \frac{c\frac{d^{M-1}z^{-n-1}}{dz^{M-1}}(\alpha)}{M!} = c \cdot \alpha^{n-M-1} n^{M-1} + \dots$$

Adding the contributions from each pole and rearranging gives the result. $\qquad\square$

By taking larger and larger R, we can refine the estimate. In fact, one can write down an infinite series expression for coefficients by including the contribution from all the singularities.

Our understanding of meromorphic functions yields a strategy to estimate the coefficients of a meromorphic function $F(z)$:

1. First, we determine the location of the poles, ρ, and their orders M_ρ.

2. Further select only the dominant of order $M = \max_\rho\{M_\rho\}$.

3. For each remaining dominant singularity we determine $C_\rho = \lim_{z\to\rho}(z-\rho)^{M_\rho} F(z)$.

4. Conclude

$$f_n \sim \left(\sum C_\rho\right)\rho^{-n} n^{M-1}.$$

More refined estimates are straightforward to determine.

4.7   The Transfer Lemma

We frequently analyze algebraic functions, with square root singularities – these are not meromorphic. What can we do?

Roughly, it is the same strategy: We look at the behaviour near the singularity. In many cases, we are able to say precise things about the coefficients from an Puiseux expansion of the function near the singularity. The intuition draws on our understanding of rational and meromorphic functions.

First, we recall that the following is known exactly for real valued α:

$$[z^n](1 - z)^\alpha = \binom{n - \alpha - 1}{n}.$$

We can estimate this using Stirling's formula:

$$\frac{n^{-\alpha-1}}{\Gamma(-\alpha)}\left(1 + \frac{\alpha(\alpha + 1)}{2n} + \dots\right).$$

We can use this to determine a first transfer result.

Theorem 4.12 (Basic Transfer). *Assume that, with the sole exception of the singularity $z = 1$, $F(z)$ is analytic in the domain*

$$\Omega = \{z \mid |z| \le 1 + v, |\arg(z - 1)| \ge \phi\}$$

for some $v > 0$ and $0 < \phi < \pi/2$. Assume further that as z tends to 1 in Ω,

$$F(z) = O(|1 - z|^\alpha)$$

for some real α. Then

$$[z^n]F(z) = O(n^{-\alpha-1}).$$

Remarkably, this proceeds by translating local behaviour of the function near 1 into information about behaviour of $F(z)$ the entire domain – as the series is given by its coefficients. The theorem is proved by a careful study of a contour integral.

We can make this more precise.

Theorem 4.13. *Assume that, with the sole exception of the singularity $z = 1$, $F(z)$ is analytic in the domain*

$$\Omega = \{z \mid |z| \le 1 + v, |\arg(z - 1)| \ge \phi\}$$

for some $v > 0$ and $0 < \phi < \pi/2$. Assume further that as z tends to 1 in Ω,

$$F(z) = (1 - z)^\alpha \log\left(\frac{1}{1-z}\right)^\beta O\left(\left(\log\frac{1}{1-z}\right)^{-1}\right)$$

for some real α and β such that $\alpha \notin \{0, 1, 2, \ldots\}$. Then

$$[z^n]F(z) = \frac{n^{-\alpha-1}}{\Gamma(-\alpha)} \log^\beta n \left(O\left(\log^{-1} n\right)\right).$$

We conclude **the second principle of coefficient asymptotics.**

Second Principle of Coefficient Asymptotics

The **nature** of the singularities determines the way the dominant exponential term in coefficients is modulated by a subexponential factor.

4.8 A General Process for Coefficient Analysis

Let us synthesize the results of this chapter to articulate a general process for coefficient analysis of combinatorial generating functions.

First, recall the two principles: The locations of the singularities dictate the exponential growth of the coefficients, and the nature of the singularities dictates the sub-exponential growth.

Consider the Catalan generating function:

$$C(z) = \frac{1 - \sqrt{1 - 4z}}{2} \qquad c_n = [z^n]C(z) = \frac{1}{n}\binom{2n-2}{n-1}.$$

The complex function $C(z)$ has a square root singularity at $z = 1/4$. The radius of convergence of the corresponding power series is $1/4$. Despite this we may still analytically continue these functions and evaluate them outside these discs. That being said, it is the behaviour close to the singularities on the radius of convergence that will be of most interest to us, because these singularities dictate the asymptotics of the coefficients. (The other singularities will give us the lower-order terms).

Given an exact expression for a coefficient, sometimes we can compute the asymptotics using some standard results (typically, Stirling's formula). In the case of c_n, this is sufficient, and we deduce $c_n \sim 4^n n^{-3/2} \sqrt{\pi}^{-1}$. Recall the notation:

$$f_n \sim g_n \Leftrightarrow \lim_{n \to \infty} \frac{f_n}{g_n} = 1.$$

Almost all the objects that we study have this same general form of asymptotics, namely

$$[z^n] f(z) = \mu^n \theta(n)$$

that is – an exponential term μ^n, modulated by a sub-exponential term $\theta(n)$. In the case of Catalan, we have $\theta(n) = O(n^{3/2})$; indeed we will always find this when we have a square-root singularity. The weakness of this approach is that we might like finer control in the error terms.

Here is an outline of how to proceed:

Locate the dominant singularity If we have the function in front of our eyes, we look for poles, logarithms, exponents. We look for the smallest real valued singularity, let us call it ρ.

Rescale the function To access the sub-exponential growth, we will use theorems that assume that the dominant singularity is at 1. This does not sacrifice generality because of the following identity:

$$[z^n] F(z) = \rho^{-n} [z^n] F(\rho z). \tag{4.6}$$

Approximate with a simpler function near the singularity We can do a Laurent expansion, or other kind of expansion near the singularity to simplify what we analyze. We must always be mindful of the region of convergence, particularly if we are evaluating series.

Apply the "transfer theorem" We can look up existing results for powers, and logarithms. Darboux's Lemma or Flajolet and Odlyzko's "transfer theorem" [FO90] states conditions under which we can use the following natural idea:

$$\lim_{z \to \rho} \frac{f(z)}{g(z)} = 1 \Rightarrow [z^n] f(z) \sim [z^n] g(z).$$

This isn't universally true: We need to verify that the function can be analytically continued beyond the singularity in a suitable region. When it is true we don't need to analyse f in detail, we just need to show that at its singularity it behaves the singularity in g, where g is simpler. Typically we will use g of the form

$$g(z) = (1-z)^\alpha \log(1-z)^\beta.$$

Their basic transfer theorem (Proposition 1) applies Stirling's formula to this decomposition:

$$[z^n](1-z)^\alpha \sim \frac{n^{-\alpha-1}}{\Gamma(-\alpha)}\left[1 + \frac{\alpha(\alpha+1)}{2n} + \cdots\right].$$

In fact, they give very precise error estimates which require more sophisticated techniques than simply Stirling's formula. Here α can be complex.

Example 4.8. One more analysis of the Catalan generating function: $C(z) = \frac{1-\sqrt{1-4z}}{2}$.

Locate the dominant singularities There is a branch point (specifically a square root singularity) when $1 - 4z = 0$, thus $z = 1/4$. There is a single dominant singularity at $\rho_C = 1/4$.

Rescale the function We reduce to a function with its dominant singularity at 1. Remark that $C\left(\frac{1}{4}z\right) = \frac{1-\sqrt{1-z}}{2}$, so, for $n > 0$,

$$[z^n]C(z) = 4^n[z^n]C\left(\frac{1}{4}z\right) = 4^n[z^n]\sqrt{1-z}. \qquad (4.7)$$

Apply the transfer theorem We apply the transfer theorem with $\alpha = -1/2$. We conclude

$$c_n \sim 4^n \frac{n^{-3/2}}{\Gamma(-1/2)},$$

with $\Gamma(-1/2) = \sqrt{\pi}$.

◄

Example 4.9. We analyse $G(z) = \frac{e^{-z/2-z^2/4}}{\sqrt{1-z}}$, which arises as an EGF of involutions.

Locate the dominant singularities The denominator has a single branch point at $\rho_G = 1$. The numerator is entire and hence is analytic at this point.

Rescale the function The singularity is already at one; there is nothing to do.

Approximate with a simpler function near the singularity The numerator is estimated with the constant term of its Taylor expansion at $z = 1$:

$$G(z) = \frac{e^{-\frac{3}{4}} - e^{-\frac{3}{4}}(z-1) + \frac{e^{-\frac{3}{4}}}{4}(z-1)^2 + \dots}{\sqrt{1-z}} = e^{-3/4}\left[\frac{1}{\sqrt{1-z}} + \sqrt{1-z} + \dots\right]$$

Apply the transfer theorem We set $\alpha = -1/2$ again, and conclude

$$g_n \sim e^{-3/4} \sim \frac{e^{-3/4}}{\sqrt{\pi n}}.$$

◄

So far, we have assumed that there is a single dominant singularity. There is distinct behaviour in the case of multiple dominant singularities, which we consider next.

4.9 Multiple Dominant Singularities

First-order asymptotics are typically given by the dominant singularity, and the others are often safely ignored as their effect on the asymptotics decays exponentially. However, interesting behaviour is possible when functions have multiple singularities on the disc of convergence: namely, periodic behaviour. Remarkably it is possible to show that certain periodic behaviour must come from multiple singularities on the disc of convergence.

An obvious example of the periodic behaviour is the function

$$F(z) = \frac{1}{1-z^3} = \sum_{k \geq 0} z^{3k}.$$

The coefficients have a periodic pattern mod 3.

By adding similar such things one can find complicated periodic patterns of whatever period you like. This can get messy when one looks at things like

$$f(z) = \frac{1}{1 - 27z^3} + \frac{1}{1 - 2z}$$

$$f_n \sim \begin{cases} 3^n & \text{if } n = 0 \mod 3 \\ 2^n & \text{if } n = \pm 1 \mod 3. \end{cases}$$

Often one can find a natural decomposition in the combinatorial model that might accompany such a periodic behaviour, and in that case, a better strategy is to handle the different cases separately at that level.

The above example has the feature that the roots are evenly spaced around two different circles: $|z| = 1/2$ or $|z| = 1/3$, which contributes to the ease of analysis. If one is unlucky, one is confronted with unevenly spaced roots, but this does not really occur in typical rational functions.

In order to formalize these notions, we define the notion of the **support** of a series. Suppose that $F(z) = \sum f_n z^n \in K[[z]]$. The support of F over K is denoted $\mathrm{Supp}(F)$ and is defined

$$\mathrm{Supp}(F) = \{n : f_n \neq 0\}. \tag{4.8}$$

A series or its coefficient sequence is said to **admit a span** d if for some r we have

$$\mathrm{Supp}(f) \subseteq r + d\mathbb{Z}_{\geq 0}. \tag{4.9}$$

The **period** of a series or sequence is the largest possible span. Note that all other spans must be divisors of the period. If the period is equal to 1 then the sequence is **aperiodic**.

It is easiest, from an analytic point of view, to work with aperiodic functions. If a function has span d, and satisfies Eq. (4.9), then there is an aperiodic analytic $G(z)$ so that $F(z) = z^r G(z^d)$. Relevant to the proof of the lemma below is the fact that if $F(z)$ has nonnegative coefficients, then so must $G(z)$.

This observation, in addition to the triangle inequality leads to a result known as the **Daffodil lemma**, which considers the behaviour of functions on a circle of convergence. For functions that diverge at

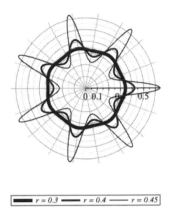

FIGURE 4.4
Radial plot of $\left|F(re^{i\theta})\right|$ for $\theta = 0..2\pi$, $r = 0.3, 0.4, 0.45$, $F(z) = \frac{1}{1-(2z)^9}$.

their dominant singularities, this connects the span and the location of singularities along the circle of convergence.

Lemma 4.14 (Daffodil lemma). *Suppose $F(z)$ is analytic in $|z| < \rho$ with series expansion around zero with nonnegative coefficients $\sum_n f_n z^n$. Assume that $F(z)$ is not a monomial and that for some non-zero, non-positive z_0 with $|z_0| < \rho$ we have*

$$|F(z_0)| = F(|z_0|).$$

Then $z_0 = Re^{i\theta}$ with θ a rational multiple of 2π. Precisely, $\frac{\theta}{2\pi} = \frac{r}{p} \in \mathbb{Q}$ with $0 < r < p$. Furthermore $F(z)$ has span p.

Figure 4.4 illustrates with $F(z) = \frac{1}{1-(2z)^9}$ and $p = 9$. This function has nine evenly-spaced singularities on the circle of convergence $|z| = 1/2$ and the divergence of $|F(z)|$ at these singularities is apparent along the "petals" of the daffodil.

Proof. Write $z_0 = Re^{i\theta_0}$. The hypothesis that $F(|z_0|) = |F(z_0)|$ implies that for all n with $f_n \neq 0$, the values $f_n R^n e^{ni\theta}$ lie along the same half-line from 0 by the triangle inequality. Since F is not a monomial, there are at least two such values, say n and m. If $f_n R^n e^{ni\theta}$ and $f_m R^m e^{mi\theta}$ are on the same half-line, then as $f_n R$ and $f_m R$ are both real, $n\theta = m\theta + 2k\pi$ for some integer k from which we conclude that θ is a rational multiple of 2π. Let us write $\theta = 2\pi\frac{r}{p}$. Indeed, as for *any* pair $n, m \in$ Supp(F),

$$(n - m)\theta \equiv 0 (\text{mod } 2\pi) \tag{4.10}$$

$$\implies (n - m)\frac{r}{p} = k \text{ for some } k \in \mathbb{Z}. \tag{4.11}$$

Hence p divides $(n - m)$ is thus a span for F. $\qquad\square$

Remarkably, we can interpret this on the level of combinatorial classes.

Proposition 4.15. *Let $\mathscr{C}$ be a constructible unlabelled class that is non-recursive. Assume that $C(z)$ has a **finite** radius of convergence ρ, $\rho < 1$. There is a $d \in \mathbb{N}$ so that the set of dominant singularities of $C(z)$ is contained in the set $\{\rho\omega^j\}$ where $\omega^d = 1$. When $\rho = 1$ it is possible that the unit circle is a natural boundary.*

Proof. By definition we obtain the class by a non-recursive construction starting from $\mathscr{E}$ and $\mathscr{Z}$ using a finite number of $+, \times, \text{SEQ}()$ constructions.

- If $\mathscr{S} = \mathscr{T} \times \mathscr{U}$ where the periods of T and U are p, q then

$$\text{Supp}(T) \subseteq a + p\mathbb{Z} \qquad\qquad \text{Supp}(U) \subseteq b + q\mathbb{Z}$$
$$\text{Supp}(S) \subseteq a + b + \gcd(p, q)\mathbb{Z}.$$

- If $\mathscr{S} = \mathscr{T} + \mathscr{U}$

$$\text{Supp}(T + U) \subseteq \min\{a, b\} + \gcd(p, q)\mathbb{Z}.$$

Some caution is required: Consider the example

$$\frac{1}{1 - z^2} + \frac{z}{1 - z^2} = \frac{1 + z}{1 - z^2} = \frac{1}{1 - z}. \tag{4.12}$$

It is an example of an aperiodic sum of two periodic functions. Cancellation can occur when there are matching periods so we must check that the resulting generating functions do not have a smaller period. This is done by checking a, b modulo various factors of p.

- If $\mathscr{S} = \text{SEQ}(\mathscr{T})$ then

$$\text{Supp}(S) \subseteq \gcd(a, p)\mathbb{Z}.$$

Since combinatorial generating functions have nonnegative coefficients, the radius of convergence of a singularity of a summand is a singularity of a finite sum. The dominant singularity is the smallest of the dominant singularities of the summands. The same is also true of a finite product.

Assume we have $S(z) = (1 - T(z))^{-1}$. The radius of convergence of $S(z)$ is ρ defined by the equation $T(\rho) = 1$. Now, if there are other singularities with the same modulus then they must satisfy $|\zeta| = 1$ and $T(\zeta) = 1$.

But this brings us back to the Daffodil lemma. Suppose $|T(z_0)| = 1 = T(|z_0|) = T(\rho)$, and so the argument of z_0 must be a rational multiple of 2π. We know that $d = \gcd(a, p)$ is a span of S, and so if $z_0 = \rho e^{i\theta}$ we can write $\theta = 2\pi j/d$, although this might not be reduced. $\square$

In Section 5.5 we consider the classic properties of combinatorial generating functions with radius of convergence 1. Some of them are quite surprising.

4.10 Saddle Point Estimation

The emphasis to this point on singularities raises an important question: How do you analyze the coefficients of function that is analytic everywhere, i.e., entire? In combinatorics, such examples arise frequently because of the exponential function, which is analytic in the entire complex plane. This does not happen naturally with ordinary generating functions, but it is common with exponential generating functions. It does become crucial to the later multivariable case, however, so let's take a look.

Lemma 4.16 (Saddle point bounds). *Let F be analytic in the disc $|z| < R$ and define $M(r) = \sup_{|z|=r} |F(z)|$. Then we have*

$$[z^n]F(z) \leq M(r)/r^n \quad \Rightarrow \quad [z^n]F(z) \leq \inf_{0<r<R} M(r)/r^n.$$

Proof. The bound follows a basic integral bound: The absolute value of an integral is bounded above by the maximum value of the integrand times the length of the contour. The proof of this result happens in two simple steps: We first apply Cauchy's coefficient formula,

Saddle point bound $[z^n]F(z) < M(r)/r^n$ is a saddle point bound if r satisfies

$$rF'(r) = nF(r),$$

and then the basic integral bound:

$$[z^n]F(z) = \frac{1}{2\pi i} \int_{|z|=r} F(z)\frac{dz^{n+1}}{dz} \leq \frac{1}{2\pi} \cdot 2\pi r \frac{M(r)}{r^n}.$$

□

When F has positive coefficients, the supremum along $|z| = r$ occurs at $z = r$ by the triangle inequality. The best possible bound of this type occurs when the appropriate derivative is zero:

$$\frac{d}{dr}\left(\frac{F(r)}{r^n}\right) = 0 \quad \Longrightarrow \quad rF'(r) = nF(r). \tag{4.13}$$

Thus, we can determine a first bound with a relatively straightforward calculus computation. The examples illustrate the process.

Example 4.10 (Factorial). Recall $e^z = \sum_{n \geq 0} \frac{z^n}{n!}$ is entire. We estimate $\frac{1}{n!}$ via a saddle point bound on $[z^n]e^z$. Equation (4.13) in this case gives

$$re^r = ne^r \quad \Longrightarrow \quad r = n. \tag{4.14}$$

The bound that this value of r gives is:

$$\frac{1}{n!} = [z^n]e^z \leq \frac{e^n}{n^n}.$$

We will prove Stirling's approximation for $n!$ in Chapter 7, which is

$$n! \sim \frac{e^n}{n^n}\frac{1}{\sqrt{2\pi n}}.$$

This is much harder to establish, and thus we are relatively content with the quality of our quick computation. ◄

It may be that an exact solution to Equation (4.13) may be difficult to find. In this case take an approximate solution – in any case a valid bound will result. It may simply be a poor bound.

4.11   Discussion

This chapter illustrates very classic aspects of series analysis. These connections strongly motivate the use of generating functions for enumeration.

For example, the convergence of a series is a fundamental topic in analysis. There are many results from the early 20th century that can be useful to the combinatorial analyst. Works of Pringshiem, Hadamard and Miller [Mil35] can be very useful to determine criteria that can apply to classes of functions to deduce preliminary estimates on the exponential growth of coefficients.

The transfer theorems are presented in detail in an important work of Flajolet and Odlyzko [FO90].

Some of the techniques from this chapter can start to address multivariable series. We have seen that **sections** of the multivariate series are important for parameter analysis. This is the subseries given by $[u^k]F(u,x)$ for some fixed, but arbitrary k. Complex analytic methods to analyze the coefficients of such a subseries are developed in much of the early literature on the multivariable asymptotics. The reader is pointed to Bender's review article [Ben74], which was further elaborated by Gao, Richmond and others [GR92].

4.12   Problems

Exercise 4.1 (Basic residue computations). 1. Prove that if f has a simple pole at z_0, and g is analytic at z_0. Then $\operatorname{res}_{z_0}(fg) = g(z_0)\operatorname{res}_{z_0}(f)$.

2. Prove that if $h(z_0) = 0$, but $h'(z_0) \neq 0$, then $1/h$ has a pole of order 1 at z_0, and $\operatorname{res}_{z_0} 1/h(z) = 1/h'(z_0)$

❏

Exercise 4.2. Using singularity analysis, find an asymptotic expression of the form $\gamma \rho^{-n} n^r$, or some other "nice" expression, for the coefficient of z^n in each of the following functions $A(z)$.

1. $A(z) = 27 \dfrac{z^3 + 4z^2 - 2}{3 - 25z + 56z^2 - 16z^3}$

2. $A(z) =$ OGF for the class of compositions such that the largest summand is less than or equal to five.

3. $A(z) = \frac{z^r}{(1-z)(1-2z)(1-3z)...(1-rz)}$. This is the OGF for set partitions into r blocks. You are estimating Stirling numbers of the second kind.

Reference: [FS09, I.4.3]

❐

Exercise 4.3. Let x_n be the nth positive root of the equation $\tan(x) = x$. For any integer $r \geq 1$ the sum $S(r) = \sum_n x_n^{-2r}$ is a rational number. The first few are $S(1) = 1/10, S(2) = 1/350, S(3) = 1/7875$. Prove these by following the steps below, and explain how to evaluate it generally.

Here is a possible template for your solution. Others are possible, and acceptable.

1. Prove that the only solutions to $\tan(x) = x$ lie on the real line.

2. Consider the function $f(z) = \frac{z^2 \tan(z)}{(z - \tan(z))}$. Find the first 6 coefficients in the Taylor series expansion around zero.

3. Now, find a second expression for $[z^n]f(z)$ using the Cauchy integral formula. Using a **careful analysis** of $f(z)$ near its singularities, express the Cauchy integral as a sum of residues.

Reference: [FS09, IV.36]

❐

Exercise 4.4. Prove the formula for Lagrange inversion. That is, start with $nT_n = [z^{n-1}]T'(z)$ and show

$$nT_n = [u^{-1}]\frac{\phi(u)^n}{u^n} = [u^{n-1}]\phi(u)^n$$

using the Cauchy integral formula. ❐

Exercise 4.5 (Zigzag-free binary words). Let $\mathcal{W}$ be the class of words over the alphabet $\{a, b\}$ said to be **zigzag-free**. That is, they contain no instance of *aba* nor *bab*. The goal is to find an asymptotic expression for the number of zigzag-free words with an equal number of *a*s and *b*s.

1. Let $W(x,y)$ be the bivariate generating function for W such that x marks the number of as, and y marks the number of bs. Prove

$$W(x,y) = \frac{1 + xy + x^2 y^2}{1 - x - y + xy - x^2 y^2}.$$

2. Let $w_{m,n}$ be the number of words in W with m as and n bs. Compute $w_{n,n}$ for $n = 0..50$, using a computer algebra program.

3. Let $F(z) = \sum w_{n,n} z^n$. Express $F(z)$ as a Cauchy integral using $W(x,y)$, and then apply the residue theorem to get a new expression for $F(z)$. Use this expression to show

$$F(z) = \sqrt{\frac{z^2 + z + 1}{z^2 - 3z + 1}}.$$

   Remark: Think of the series $W(z,t)$ as a Laurent expansion in the t variable parametrized by z. Consider the substitution $z \mapsto z/t$ into this expression, and then figure out how to get $F(z)$. To compute this integral you will need to determine the poles, which will be functions of z, and it is sufficient to consider the poles $\alpha(z)$ which tend to 0 as z tends to 0. You can just assume this last fact if you aren't able to prove why it is so.

4. Use your expression to find an asymptotic estimate a_n for $w_{n,n}$. Plot $w_{n,n}/a_n$ for $n = 1..50$ to see how close you are.

Reference: [PW08]

❐

Exercise 4.6. Show that $\sqrt{z}$ is singular at $z = 0$ by proving that the limit in Eq. (4.1) is not defined in a neighbourhood of 0. ❐

5

Parallel Taxonomies

CONTENTS

5.1   Rational Functions 112
5.2   Algebraic Functions 113
5.3   D-finite Functions 115
       5.3.1   Closure Properties 116
       5.3.2   Is It or Isn't It? 117
       5.3.3   G-functions 120
       5.3.4   Combinatorial Classes with D-finite Generating
                Functions 120
5.4   Differentiably Algebraic Functions 121
5.5   Classification Dichotomies 123
5.6   The Classification of Small Step Lattice Path Models 124
       5.6.1   A Simple Recursion 125
       5.6.2   Models with D-finite Generating Functions 127
       5.6.3   Models with Non-D-finite Generating Functions 129
5.7   Groups and the Co-growth Problem 130
       5.7.1   Excursions on Cayley Graphs 131
       5.7.2   Amenability vs. D-finiteness 132
5.8   Discussion 133
5.9   Problems 136

We are now familiar with several families of combinatorial classes and generating functions. We can ask if there are meaningful parallels in their complexity. We are not a priori assured that that this is a meaningful line of study – some pathological languages in complexity theory have very simple generating functions! In this chapter, we first review a taxonomy of series and analytic functions and consider the induced classification on combinatorial classes. We can compare this to combinatorially defined classifications, the origin of which are questions on classical computational complexity. The function classes we consider are organized in Figure 5.1. The containment is strict. For each class, we describe some useful analytic properties, closure properties and characterizations of the combinatorial classes whose

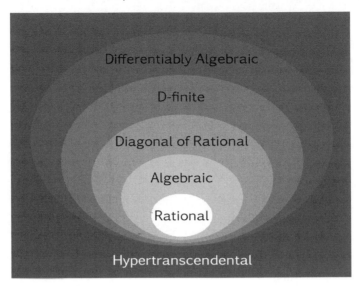

FIGURE 5.1
Venn diagram of function classes.

generating functions lie in that class. The classes are by and large defined by functional equations of very classical types, and there is a
vast historical literature. Even when we cannot solve these equations,
we can determine useful information such as the location and nature
of the dominant singularity. This is often sufficient for classification
purposes: information on the asymptotic growth of a function's Taylor
series coefficient sequence can be quite informative.

5.1   Rational Functions

A series $F \in K[[\mathbf{z}]]$ is rational over K if it satisfies an equation of
the form $Q(\mathbf{z})F(\mathbf{z}) = P(\mathbf{z})$ for polynomials $P, Q \in K[\mathbf{z}]$. We write $F \in$
$K(\mathbf{z})$. It is furthermore qualified to be $\mathbb{N}$-rational if it is contained in
the smallest set of rational functions containing $1, z_1, \ldots, z_d$ which is
closed under addition, multiplication and quasi-inverse.

  Theorem 4.10 contained very precise information about the location of the singularities, and the form of the asymptotics of rational
function coefficients. It turns out that $\mathbb{N}$-rational functions have additional analytic structure that can be exploited.

Theorem 5.1 (Berstel, 1971). *If $F(z)$ is $\mathbb{N}$-rational with radius of convergence ρ then ρ is a pole of $F(z)$. Furthermore if ρ' is a pole of $F(z)$ with $|\rho'| = \rho$, then ρ'/ρ is a (integral) root of unity.*

If $F(z)$ is $\mathbb{N}$-rational then it can be expressed as a sum of rational functions with a single (positive) pole on the circle of convergence.

Compare this to Proposition 4.15.

Example 5.1. The rational function

$$F(z) = \frac{z + 5z^2}{1 + z - 5z^2 - 125z^3} = z + 4z^2 + z^3 + 144z^4 + \ldots \quad \text{(OEIS A094423)}$$

is in $\mathbb{N}[[z]]$ but is *not* $\mathbb{N}$-rational by this theorem: The roots of the denominator labelled, respectively, ρ_1, ρ_2, ρ_3 are

$$1/5, \; -\frac{3}{25} - \frac{4}{25}i, \; -\frac{3}{25} + \frac{4}{25}i.$$

Now

$$\frac{\rho_2}{\rho_1} = -\frac{3}{5} - \frac{4i}{5}$$

is not a root of unity, since $\arg \rho_2/\rho_1 = \arctan(3/4)$ is not a rational multiple of π. ◄

A combinatorial class with an S-regular specification has an $\mathbb{N}$-rational generating function. It remains open to find a natural combinatorial source for rational generating functions that are not $\mathbb{N}$-rational.

5.2 Algebraic Functions

A series $F(\mathbf{z})$ is algebraic over K if there is a non-trivial polynomial $P \in K[\mathbf{z}, y]$, such that $F(\mathbf{z})$ satisfies the polynomial equation $P(\mathbf{z}, F) = 0$. Rational functions are algebraic. Some authors write $F \in K\langle\mathbf{z}\rangle$ in this case. If we are given the polynomial equation that a univariate algebraic function satisfies, we can determine the dominant singularities of the solution. Suppose that $F(z)$ is analytic at the origin and satisfies $P(z, F(z)) = 0$ for some $P(x, y) \in \mathbb{Q}[x][y]$. By the implicit function theorem, a singularity ρ of $F(z)$ is contained in either the set of roots of the leading term of $P(x,y)$ as a function of y, or the discriminant of $P(x,y)$ with respect to y. The latter is the resultant of P and $\partial P/\partial y$ up to a constant.

To actually determine which of these (finite in number) possible points are actually singularities, one can follow a numerical algorithm which tests to see for which points the local behaviour is singular. This amounts to studying the local Puiseux expansion. The line of study leads to a characterization of possible asymptotic behaviour of the coefficients of an algebraic series.

Theorem 5.2. *Let $F(z) = \sum_{n\geq 0} f_n z^n$ be an algebraic function over $\mathbb{Q}$, which is analytic at the origin. Suppose that the dominant singularities are $\rho\omega_1, \ldots, \rho\omega_k$ with $|\omega| = 1$ for all j. Then*

$$f_n = \frac{\rho^n}{\Gamma(\alpha + 1)} \left(\sum_{j=0}^{k} C_j \omega_j^n n^\alpha + O(n^{\alpha'}) \right), \tag{5.1}$$

where $\alpha \in \mathbb{Q} \setminus \{-1, -2, \ldots\}$, $\alpha' < \alpha$ and ρ, the ω_j, and the C_j, are all algebraic numbers.

This asymptotic result can be employed as a useful criterion for transcendence. For example, the number of simple excursions e_n in the entire plane grows asymptotically like $e_n \sim 4^n (\pi n)^{-1}$, hence the generating function $\sum_{n\geq 0} e_n z^n$ is not algebraic over $\mathbb{Q}$.

Recall we saw earlier that if a combinatorial class was specified by an algebraic system, it had an algebraic generating function. In analogy with the $\mathbb{N}$-rational series, we ask: Is there a natural source for combinatorial classes with an algebraic generating functions that cannot be written using an algebraic specification? We return to this question in the discussion.

Example 5.2. A transcendental OGF implies that a class has no algebraic specification. Consider this example from [Fla87]. We can frame the result into a transcendency criterion: If

$$[z^n]F(z) \sim C\mu^n n^s, \quad s \notin \mathbb{Q} \setminus \{-1, -2, \ldots\},$$

then F is transcendental. Consider:

$$\mathscr{C} = \{u \in \{a, b, c\}^* \mid |u|_a \neq |u|_b \text{ or } |u|_a \neq |u|_c\}.$$

We can relate its OGF to some known series to deduce information:

$$\{a,b,c\}^* \setminus \mathscr{C} = \{u \in \{a,b,c\} \mid |u|_a = |u|_b = |u|_c\}$$

$$\implies 3^n - c_n = \binom{3n}{n,n,n}$$

$$\implies \underbrace{\sum 3^n z^n - \sum c_n z^n}_{\text{rational}} = \underbrace{\sum \binom{3n}{n,n,n} z^n}_{\sim C\, 27^n\, n^{-1}}.$$

$$\underbrace{\phantom{\sum \binom{3n}{n,n,n} z^n}}_{\text{transcendental}}$$

We conclude that $\sum c_n z^n$ transcendental since the difference of two algebraic functions is algebraic.　◂

5.3 D-finite Functions

Algebraic functions are contained in the set of **differentiably finite functions**. This class is also known as the D-finite functions, and it also corresponds to holonomic functions. These functions were introduced to combinatorialists in a 1980 paper of Richard Stanley [Sta80], and they have been central to discussion in enumerative combinatorics ever since. D-finite generating functions are highly correlated with structure in the underlying combinatorial class.

Everything is non-holonomic unless it is holonomic by design.
FLAJOLET, GERHOLD AND SALVY

Let K be a field, and consider a power series over K, $F(z) = \sum f_n z^n \in K[[z]]$. We say that $F(z)$ is **differentiably finite of order** r, or, **D-finite** with respect to the variable z if there exist $r+1$ polynomials $p_0(z)\ldots,p_{d-1}(z) \in K[z]$ with $p_d(z)$ non-zero so that

$$p_r(z)\frac{d^r F(z)}{dz^r} + \cdots + p_1(z)\frac{dF(z)}{dz} + p_0(z)F(z) = 0. \qquad (5.2)$$

The prototypical example is the exponential series $e^z = \sum \frac{z^n}{n!}$, which is D-finite of order 1, since $\frac{de^z}{dz} - e^z = 0$. To prove that a series is D-finite, it is equivalent to show that the vector space over $K[z]$ generated by the partial derivatives is finite dimensional. This is an easier criterion to generalize into higher dimensions. A multivariate series $F(\mathbf{z})$ is said to be D-finite if the vector space generated by all of its partial derivatives is finite dimensional. This is equivalent to a system of d linear differential equations each with respect to a single variable, with coefficients in $K[\mathbf{z}]$.

The coefficients of univariate D-finite functions are easily characterized. A sequence $a(0), a(1), \ldots$ over K is said to be **P-recursive of order** k if there are polynomials $q_0(z), \ldots, q_k(z) \in K[n]$ such that

$$q_0(n)a(n) + q_1(n)a(n+1) + \cdots + q_k(n)a(n+k) = 0. \tag{5.3}$$

The coefficients of a univariate D-finite function form a P-recursive sequence, and it is straightforward to translate between the differential equation satisfied by the series, and the recurrence satisfied by the coefficients.

A D-finite function can be naturally encoded by a differential equation or system of differential equations with uniqueness assured by a finite set of initial conditions. The differential equation is a useful data structure and many of the closure properties can be made effective by working on the level of the differential equation. More than that, is that we are assured a finite representation of the generating function, an appealing property in general, but additionally it is one which captures many aspects of its complexity.

5.3.1   Closure Properties

Algebraic functions are D-finite. Moreover, for an algebraic function $F(z)$ that is the solution of a polynomial equation $P(z, F(z)) = 0$ (where $P(x, y)$ has degree d in y), the dimension of the space spanned by the derivatives $\frac{d^k}{dz^k}F(z)$ is bounded by d. Properties of D-finite series extend to be properties of D-finite functions. There are several closure properties of interest to the combinatorialist.

Theorem 5.3. *Suppose that F and G are D-finite functions over K. Then the following are also D-finite over K:*

1. $F + G$;

2. FG;

3. $\frac{dF(z)}{dz}$;

4. $\int_0^z F(s)\,ds$;

5. If F and G have series expansions $F(z) = \sum_{n\geq0} f_n z^n$ and $G(z) = \sum_{n\geq0} g_n z^n$, then $\sum_{n\geq0} f_n g_n z^n$, the **Hadamard product** of F and G;

Additionally,

1. $F(G(z))$ *is D-finite under the additional hypotheses that* $G(z)$ *is algebraic;*

2. F *and* $1/F$ *are simultaneously D-finite if and only if* F'/F *is algebraic;*

3. F *and* e^F *are simultaneously D-finite if and only if* F'/F *is algebraic.*

Roughly, it is the closure of the set of D-finite functions under the Hadamard product that is the property assuring their relevance to the combinatorialist. It means that the diagonal of a D-finite multivariable function is D-finite, and also that D-finite functions are closed under the Borel transform and its inverse. The Borel transform maps $\sum z^n$ to $\sum \frac{z^n}{n!}$, hence the OGF of a sequence is D-finite, if and only if the EGF is D-finite. This, and the explicit connection to P-recursive sequences, suggests that D-finiteness is a fundamental property of a combinatorial counting sequence. It must be emphasized that Item 5 in the above theorem is a decidedly nontrivial result. Proved by Lipshitz in 1988 [Lip88], it has profound consequences such as Theorem 5.4 below. Furthermore it is the most difficult to make effective.

Theorem 5.4. *The diagonal of a multivariate D-finite series is D-finite in the variables that remain.*

5.3.2 Is It or Isn't It?

For your reference, the following criteria are the most common used in arguments to conclude that a function is not D-finite.

Criterion 1

A D-finite function has a finite number of singularities. They are a subset of the roots of the polynomial of leading order.

Criterion 2

The asymptotic growth of a D-finite function satisfies a particular template. (See Theorem 5.6.)

Criterion 3

For any dominant singularity of a D-finite function, there exists a derivative so that the value of the derivative near the singularity is unbounded (Theorem 5.7).

Criterion 4

F and e^F are simultaneously D-finite if and only if F'/F is algebraic.

Example 5.3. We can show that the series $F(z) = \sum_{n=0}^{\infty} z^{2^n}$ is not D-finite in a couple of different ways. The space between terms is unbounded. Thus, there is no fixed-length recurrence for the coefficients. If the coefficients do not form a P-recursive sequence, then the function is not D-finite.

For a second argument, we note that it has a natural boundary at the unit circle: if ρ is a singularity, then so is τ such that $\tau^2 = \rho$. As $F(z)$ diverges at 1, $\rho = 1$ is a singularity, implying that -1 is also a singularity implying that $i, -i$ are singularities, ...

We can also see that F satisfies the functional equation:

$$F(z) = z + F(z^2).$$

This is a **Mahler**-type equation. A Mahlerian operator maps a function $F(z)$ to $F(z^k)$ for some k. A series with a non-zero, finite radius of convergence that satisfies a Mahlerian equation is not D-finite [Rub89]. ◂

Example 5.4. In the multivariate case, one must be careful with the arguments about singularities accumulating. For example, consider

$$F(x,y) = \sum_{n \geq 1} \frac{x^n}{1 - nx} y^n.$$

The coefficient of y^n is singular at $x = 1/n$. The set of singularities of the coefficients is infinite! However, this is still a D-finite function of y, as shown in [Rec06a]:

$$xy^2(1 - xy)\frac{\partial^2 F}{\partial y^2} - y(1 - xy + z^2 y)\left(\frac{\partial F}{\partial y} + F\right) = 0.$$

◂

The set of singularities of the coefficients of a D-finite series may be infinite, and have accumulation points, but it may not have an infinite number of accumulation points.

Theorem 5.5. *Let*

$$F(x,y) = \sum_{n \geq 1} H_n(z) y^n$$

be a D-finite series in y with coefficients $H_n(z)$ that are rational functions of x. For $n \geq 0$, let S_n be the set of poles of $H_n(z)$, and let $S = \cup S_n$. Then S has only a finite number of accumulation points.

Suppose that $F(z) \in \mathbb{R}[[z]]$ is D-finite and, in particular, is a solution the following linear differential equation:

$$p_r(z) \frac{d^r F(z)}{dz^r} + \cdots + p_0(z) F(z) = 0. \tag{5.4}$$

We can use this information to determine information about the coefficients of $F(z)$, particularly if we already have some information about the bounds of the growth.

The singularities of $F(z)$ are contained in set $\{z \in \mathbb{C} : p_r(z) = 0\}$, which is finite. There may be points in that set which are not singularities, and this can make a systematic study difficult. However, using the fact that our solution of interest is a combinatorial generating function narrows the candidates, since by Pringsheim's lemma there must be a dominant singularity on the positive real line. If we have other bounds on the exponential growth of the generating function, this can also limit the modulus of the solutions.

The next step, in analogy to the algebraic case, is to find an initial series development around the singularity. Several computer algebra systems possess the ability to do this systematically.

Theorem 5.6. *Suppose that $F(z) \in \mathbb{Z}[[z]]$ is a D-finite series, analytic at the origin. Then the dominant asymptotics of the coefficient sequence f_n is given by a finite sum of terms of the form $\gamma n^\alpha (\log n)^\ell \mu^n$, where γ is a constant, $\alpha \in \mathbb{Q}$, $\ell \in \mathbb{N}$ and μ is algebraic.*

This theorem is a corollary of a powerful result of Chudnovsky and Chudnovsky [CC85]. It is used frequently to demonstrate that a combinatorial generating function is not D-finite either by showing that α is transcendental or that α is irrational.

5.3.3 G-functions

An analytic power series $F(z) = \sum f_n z^n$ in $\mathbb{Q}[[z]]$ is said to be a **G-function** if it is D-finite, and there exists a constant $C > 0$ such that for all n, both f_n and the least common denominator of $|f_0|, \ldots, |f_n|$ are bounded by C^n. Remark this holds for ordinary generating functions with finite, non-zero radius of convergence. G-functions are a well-studied subclass with many classic results useful to the combinatorialist: Indeed, the "G" is for geometric as they are seen to generalize geometric series, which are inherently combinatorial.

If F is a G-function then it is annihilated by a Fuchsian differential equation and consequently the singularities of F are all regular. Following work from Chudnovsky and Chudnovsky[CC85], we have that if ρ is a singularity of F then under some technical conditions near ρ we have

$$F(z) = \sum_{i=1}^{s} \sum_{k=0}^{k_j} (\rho - z)^{\lambda_i} (\log(\rho - z))^k f_{i,k}(\rho - z),$$

with $\lambda_1, \ldots, \lambda_d$ rational numbers such that $\lambda_i - \lambda_j \notin \mathbb{Z}$ for $i \neq j$.

As a corollary of the form of G-series, the behaviour of their derivatives is equally well understood, providing another criterion for non-D-finiteness, which is applicable to many series coming from combinatorics.

Theorem 5.7. *Let $F(z) \in \mathbb{N}[[z]]$ be a power series with nonnegative integer coefficients and suppose that $F(z)$ has radius of convergence $R > 0$. If the $\frac{d^j}{dz^j F}(z)$ absolutely converges on the closed disc of radius R for every $j \geq 1$ then F is not D-finite.*

Compare with Theorem 5.11.

5.3.4 Combinatorial Classes with D-finite Generating Functions

Stanley's introductory article [Sta80] raised many questions about which combinatorial classes have a D-finite generating function. In Part I we have seen the derived classes and the reflectable walks. This represents a large number of classes with D-finite generating functions, indeed a great many if you consider combinatorial classes in bijection with such classes. The Robinson-Schensted correspondence is a famous non-trivial map between pairs of standard Young tableaux of the same shape and permutations. It figures centrally to many bijections between classes of reflectable walks and the world of permutations. Knowing this, it is far less surprising that many permutation classes are D-finite, as often the proof is a reduction to a reflectable walk. Baxter permutations and involutions with bounded maximal increasing subsequences are two notable such families of permutations. Gessel [Ges90] used symmetric functions to deduce the D-finiteness of some generating functions that are **not** G-series (the OGF have radius of convergence 0) including notably the generating functions for labelled k-regular graphs[1] His construction builds a derived class from the family of all labelled graphs.

Is there a natural family of combinatorial classes whose ordinary generating function is transcendental D-finite but the class is not a derived class? Is there a class with a D-finite generating function with a finite radius of convergence that are not expressible as a diagonal of a rational function? If such a class can be found, it would might lead to a clear answer to the following question of Christol.

Question 5.1 (Christol, 1990 [Chr90]). Can every G-series be expressed as a diagonal of a rational series?

5.4 Differentiably Algebraic Functions

Beyond the set of D-finite functions is a bit of a wilderness. There are several natural ways to generalize D-finite functions. A class that is already well studied is the set of differentiably algebraic functions. A series $F(z) \in K[[z]]$ is differentiably algebraic if there exists a polynomial $P \in K[x]$ of degree $k + 1$ such that $P(z, F(z), F'(z), \ldots, \frac{d^k F}{dz^k}) = 0$. Clearly this class of function contains the set of D-finite functions.

This function class has closure properties relevant to combinatorics.

Theorem 5.8. *Suppose that $F(z) \in K[[z]]$ and $G(z) \in K[[z]]$ are differentiably algebraic. Then so are $F + G$, FG, $F(G)$ and $1/F$.*

---

[1] These are graphs labelled with 1 to n such that every vertex has degree k. The result holds for different types of graphs including simple graphs, multigraphs and graphs with loops.

By this closure theorem, $F(z) = e^{e^{z-1}}$ is differentiably algebraic. We conclude that the class of differentiably algebraic functions is strictly larger than the class of D-finite functions.

The proofs of the different properties stated in Theorem 5.8 are classic and date to the 19th century. Differentiably algebraic functions came up in the study of the elliptic theta functions, which are the elliptic analogues of the exponential function. Similar to D-finite functions, differentiably algebraic functions cannot have unbounded distance between non-zero terms.

A **hypertranscendental** function does not satisfy any algebraic differential equation. (For example, the Riemann zeta function). Mahler obtained many general results about hypertranscendence.

There are criteria for establishing that a series is not differentiably algebraic that rely on facts of q-difference equations. Ishizaki proved a theorem on differential transcendency of meromorphic functions satisfying linear q-difference equations of first order, and Ogawara adapted the result to say something about formal Laurent series.

Theorem 5.9. *Let q be a non-zero complex number with $|q| \neq 1$. Suppose that a meromorphic function $F(z)$ on $\mathbb{C}$ satisfies a linear q difference equation:*

$$F(qz) = A(z)F(z) + B(z) \tag{5.5}$$

of first order for some rational functions $A(z), B(z) \in \mathbb{C}$. Then $F(z)$ does not satisfy any nontrivial algebraic differential equation over $\mathbb{C}(z)$ unless it represents a rational function.

Ogawara proved that a formal Laurent series satisfying a rational linear q-difference equation of first order does not satisfy any nontrivial algebraic differential equation over the rational function field unless it actually is a rational function.

Theorem 5.10. *[Oga14] Let K be an algebraically closed field of characteristic 0, and $q \in K$ a non-zero element that is not a root unity. Suppose that a Laurent series $F(z) \in K((z))$ satisfies the following two conditions:*

1. *$F(qz) = A(z)F(z) + B(z)$, for rational functions $A(z)$ and $B(z)$,*

2. *$F(z)$ is differentiably algebraic over $K(z)$.*

Then $F(z)$ is a rational function in $K(z)$.

5.5 Classification Dichotomies

Formal series arising from combinatorics have the particular property that the coefficients are all natural numbers. Such series exist in some powerful dichotomies. The earliest of these date back well over 100 years and are attributed to Fatou.

Theorem 5.11 (Fatou [Fat06]). *Suppose $F(z)$ is a function with a series expansion $F(z) = \sum_{n \geq 0} f_n z^n$ such that $f_n \in \mathbb{Z}$. If $F(z)$ converges inside the unit disc, then it is either a rational function or transcendental over $\mathbb{Q}(z)$.*

Theorem 5.12 (Pólya Carlson theorem). *A series $F \in \mathbb{Z}[[x]]$ that converges inside the unit disc is either rational or has the unit circle as a natural boundary.*

We use this as follows: If a series converges inside the unit disc and is not rational, then it is also not D-finite.

Recently Bell and Chen determined multivariable versions of some classic results. They reference multivariable convergence, which is defined in the next chapter.

Theorem 5.13 (Bell, Chen [BC17]). *A D-finite multivariate power series with integer coefficients that converges on the unit polydisk is rational.*

Van der Poorten and Shparlinski proved the following [vdPS96].

Theorem 5.14. *Consider the map $f(n) : \mathbb{N} \to \mathcal{N}$ such that $\mathcal{N}$ is a finite subset of $\mathbb{Q}$. If the generating function $F(z) = \sum_{n \geq 0} f(n) z^n$ is D-finite, then in fact it is rational.*

This has a multivariable version by Bell and Chen [BC17].

Theorem 5.15. *Let $f(n_1, ..., n_d) : \mathbb{N}^d \to \mathcal{N} \subset \mathbb{Q}$ where $\{|\mathbf{n}| \mid \mathbf{n} \in \mathcal{N}\}$ is a finite subset of $\mathbb{Q}$. If the generating function $\sum_{\mathbf{n} \in \mathbb{N}^d} f(\mathbf{n}) \mathbf{z}^{\mathbf{n}}$ is D-finite, then it is rational.*

It is quite rare for a function to satisfy several types of equations. Indeed, when this happens, we can often show that the function is either rational or hypertranscendental [SS17].

Example 5.5. The classic example is the Γ function: $\Gamma(z + 1) = z\Gamma(z)$. Very roughly, it satisfies a shift equation, and is not rational since it has an infinite number of poles. Thus, it is hypertranscendental. ◄

Classification of Lattice Walks

Determine combinatorial criteria on the step set $\mathscr{S}$ that determine the nature of the generating function $Q^{\mathscr{S}}(x, y, z)$ marking length and endpoint.

Example 5.6. We return to the series $F(z) = \sum_{n=0}^{\infty} z^{2^n}$. It is easy to test that F satisfies the functional equation:

$$F(z) = z + F(z^2).$$

This is a **Mahler**-type equation. A Mahlerian operator maps a function $f(z)$ to $f(z^k)$ for some positive integer k. A series with a non-zero, finite radius of convergence satisfies a Maherlian equation is either rational or hypertranscendental.   ◀

5.6   The Classification of Small Step Lattice Path Models

Lattice walk models are easy to wrap our head around, and they demonstrate some features of generating function analysis. We can illustrate many of the classification results in this chapter on two-dimensional lattice models restricted to the first quadrant $\mathbb{N}^2$. In fact, the taxonomy of models of two-dimensional walks restricted to the first quadrant is a surprisingly rich activity.

We make a finite problem by considering only those models that take "small" steps, that is models with a step sets $\mathscr{S} \subset \{0, \pm 1\}^2$. That is, each step to one of the 8 nearest integer lattice points, and thus, there are $2^8 - 1$ lattice models definable with small steps. Step sets that take only steps in a negative direction only contain the trivial walk. Because of the symmetry of the domain, we need only count one of a model and its image across the line $x = y$. So far, we have seen several examples of such combinatorial families, notably the walks in the Weyl chambers in Chapter 3. The **complete generating function** for the family of lattice walks with steps from $\mathscr{S}$ that start at the origin is the series

$$Q^{\mathscr{S}}(x,y,z) := \sum_{n\geq 0}\sum_{(k,\ell)\in\mathbb{N}^2} \text{walk}^{\mathscr{S}}((0,0)\xrightarrow{n}(k,\ell))\,x^k y^\ell z^n. \qquad (5.6)$$

Typically, the step set is clear, and we do not include it when we reference the function.

Given a model, $\mathscr{S}$, can we determine the nature of its generating function? Is there any correlation between the nature of the series and combinatorial properties of the model? Many combinatorial families are in bijection with lattice models, and so this question has implications across combinatorics.

5.6.1   A Simple Recursion

Now, a walk is either a walk of length 0, or a walk followed by a valid step. This straightforward decomposition has a natural translation into generating functions. A property specific to models of walks in the quarter plane is that the generating function for walks that end on a boundary can be obtained from an evaluation of $Q(x,y,z)$. For example, the family of walks that begin and end at the origin have generating function

$$\sum_{n\geq 0}\text{walk}^{\mathscr{S}}((0,0)\xrightarrow{n}(0,0))\,z^n = Q(0,0,z).$$

The walks that end on the y-axis have generating function $\sum_{n,k\geq 0}\text{walk}^{\mathscr{S}}((0,0)\xrightarrow{n}(k,0))\,x^k z^n = Q(x,0,z)$. This is a particularly convenient representation as the combinatorial decomposition translates directly into a recurrence for the complete generating function. Let $S(x,y) = \sum_{(s_1,s_2)\in\mathscr{S}} x^{s_1} y^{s_2}$.

$$Q(x,y,z) = 1 + zS(x,y)Q(x,y,z)$$
$$- z\sum_{(s_1,-1)\in\mathscr{S}} x^{s_1}y^{-1}Q(x,0,z) - z\sum_{(-1,s_2)\in\mathscr{S}} x^{-1}y^{s_2}Q(0,y,z)$$
$$+ z\delta_{-1,-1}x^{-1}y^{-1}Q(0,0,z)$$

which we reorganize into

$$K(x,y,z)Q(x,y,z) = xy - R(x,z) - S(y,z) + z\delta_{-1,-1}Q(0,0,z) \qquad (5.7)$$

with

$$K(x,y,z) := xy\,(1-zS(x,y))$$
$$R(x,z) := K(x,0,z)Q(x,0,z), \quad \text{and} \quad S(y,z) := K(0,y,z)Q(0,y,z)$$
$$\delta_{-1,-1} = 1 \text{ if } (-1,-1) \in \mathscr{S}, 0 \text{ otherwise.}$$

Many enumeration strategies start with Eq. (5.7), which is known as a **kernel equation**, and then differ in their subsequent manipulations. Since our steps are small, $K(x,y,z)$ is quadratic. This is a key feature in much of the analysis. Considering larger steps is considerably more complicated.

Example 5.7. The functional equation in the case of $\mathscr{S} = \{(1,0),(1,1),(0,1)\}$ is straightforward since there are no negative steps:

$$Q(x,y,z) = 1 + zS(x,y)Q(x,y,z) \implies Q(x,y,z) = \frac{1}{1-zS(x,y)}.$$

This is a rational function. ◄

Example 5.8. Consider a model with no downward steps, such as $\mathscr{S} = $ ↘. The functional equation can be solved in the manner of Section 2.7 using an explicit solution $Y(x,z)$ to the algebraic equation $K(x,Y(x)) = 0$. As K is quadratic, we can use the quadratic equation to solve this. One of the solutions is analytic at 0. As $S(y,z) = 0$ in this case, Eq. (5.7) becomes

$$0 = xY(x,z) - R(x,z) \implies Q(x,0,z) = \frac{xY(x,z)}{K(x,0,z)},$$

which also solves for $Q(x,y,z)$ since:

$$Q(x,y,z) = \frac{xy - Q(x,0,z)}{K(x,y,z)}.$$

The generating function is algebraic. Indeed, any model with steps that interact only with a single boundary will have an algebraic generating function. Indeed we can deduce a grammar that specifies the class. ◄

Given these analyses, we focus on models that interact with both boundaries. After trivial walks and duplicates are eliminated, there are 79 distinct, interesting models left. The classification of the complete generating function of the 79 models was completed after a nearly 20-year study. The results are tabulated in Table 5.1.

TABLE 5.1

The Generating Function Taxonomy for Quarter-plane Lattice Path Model-generating Function $Q^{\mathscr{S}}(x,y,z)$

Class	Count	Step Set $\mathscr{S}$
Algebraic	4	⚹ ⊢ ⊣ ⚹
D-finite	19	+ ✕ ✳ ✴ ⅄ ⅄ ⅄ ✴ ✕ ✳
		⼊ ✳ ✕ ✴ ⼊ ⼊ ✴ ⼊ ⼊
D-algebraic	9	⼊ ⼊ ⼊ ⼊ ⼊ ⼊ ⼊ + ⼊
Hyper transcendental	42	✕ ⅄ ⅄ ✕ ✳ ✳ ✳ ✳ ✳ ✳
		⅄ ⅄ ⅄ ⅄ ⅄ ⅄ ⅄ ⅄ ✳
		⼊ ⼊ ⼊ ⼊ ⼊ ⼊ ⼊ + +
		⼊ ⼊ ✳ ⼊ + ⼊ ⼊ ⼊ ⼊
		⼊ ⼊ ⼊ ⼊ ⼊
	5	⅄ ⅄ ⅄ ⅄ ⅄

The four nontrivial algebraic models ⚹ ⊢ ⊣ ⚹ are algebraic, and we can use arguments on the asymptotics of the counting sequence to prove that they cannot be specified by an algebraic grammar.

5.6.2 Models with D-finite Generating Functions

We have seen that models that arise as walks in Weyl chambers have D-finite generating functions, as they can be expressed as diagonals of rational functions. Most of the remaining transcendental models are captured by the following theorem.

Theorem 5.16. *Let $\mathscr{S}$ be a finite subset of $\mathbb{Z} \times \{\pm 1, 0\}$ that is symmetric with respect to the x-axis. Let $Q(x,y,z)$ be the complete generating function for walks that start from $(0,0)$, take their steps in $\mathscr{S}$ and stay in the first quadrant. Then, $Q(x,y,z)$ is D-finite in each of its variables.*

Proof. The symmetry condition implies that $S(1/x, y) = S(x, y)$. This motivates a variable substitution of $x \mapsto 1/x$ in the kernel equation:[2]

$$K(x,y,z)Q(x,y,z) = xy - R(x,z) - S(y,z) + z\delta_{-1,-1}Q(0,0,z) \quad (5.8)$$

$$K\left(\frac{1}{x},y,z\right)Q\left(\frac{1}{x},y,z\right) = \frac{y}{x} - R\left(\frac{1}{x},z\right) - S(y,z) + z\delta_{-1,-1}Q(0,0,z). \quad (5.9)$$

---

[2]Care must be taken that this makes sense and that we track the ring that we are working in. In this case, $K[x,x^{-1},y,y^{-1}][[z]]$, so the substitution is valid.

If $Y(x,z)$ is a solution to $K(x, Y(x,z)) = 0$ analytic at the origin, then the symmetry implies $K(1/x, Y(x,z)) = 0$ as well. Setting $y = Y(x,z)$ in Eq. (5.8), and taking the difference gives:

$$0 = \left(x - \frac{1}{x}\right) Y(x,z) - R(x,z) + R\left(\frac{1}{x}, z\right) \qquad (5.10)$$

$$\Longrightarrow R(x,z) - R(0,z) = [x^{>0}]\left(\frac{1}{x} - x\right) Y(x,z)^3 \qquad (5.11)$$

since $R(x,z)$ has only nonnegative powers of x (and so $R(1/x,z)$ has only nonpositive powers of x). We conclude that $R(x,z) - R(0,z)$ is a D-finite function. Now, to make the final argument, we rewrite the kernel equation using this term. It eliminates the need for an inclusion exclusion:

$$K(x,y,z)Q(x,y,z) = xy - (R(x,z) - R(0,z)) - S(y,z). \qquad (5.12)$$

Let X be the solution to the kernel $K(X(y), y) = 0$:

$$0 = X(y)y - (R(X(y), z) - R(0,z)) - S(y,z).$$

Then

$$K(x,y,z)Q(x,y,z) - 0$$
$$= xy - (R(x,z) - R(0,z)) - S(y,z)$$
$$\quad - X(y)y + (R(X(y), z) - R(0,z)) + S(y,z)$$
$$= xy - X(y)y - (R(x,z) - R(0,z)) + (R(X(y), z) - R(0,z)).$$

We express $Q(x,y,z)$ as a sum of algebraic functions and D-finite functions evaluated at an algebraic function, therefore it is D-finite. $\qquad \square$

Example 5.9. Consider the diagonal walks with step set $\mathscr{S} = \times$. The kernel is $K(x,y,z) = xy(1 - z(xy + x/y + y/x + 1/xy))$. The solution uses two roots of the kernel: $0 = K(x, Y(x,z)) = K(X(y,z), y)$, both must be series solutions; for example

$$Y(x,z) = \frac{x - \sqrt{-4x^4z^2 - 8x^2z^2 + x^2 - 4z^2}}{2z(x^2 + 1)}$$

$$= \left(x + \frac{1}{x}\right)z + \left(x^3 + 3x + 3/x + 1/x^3\right)z^3 + \dots$$

and

$$X(y,z) = -\frac{-y + \sqrt{-4y^4z^2 - 8y^2z^2 + y^2 - 4z^2}}{(2z(y^2 + 1))}.$$

---

[3] Here keep the powers of x in the extraction of positive terms.

We note that $Y(x,z)$ is invariant under $x \mapsto 1/x$. We use

$$RR(x,z) := R(x,z) - R(0,z)$$

$$= [x^{>0}]\left(\frac{1}{x} - x\right)Y(x,z)$$

$$= xz + \left(x^3 + 3x\right)z^3 + \left(2x^5 + 10x^3 + 20x\right)z^5 + O(z^7).$$

The final result is a closed form expression for the complete generating function

$$Q^{\times}(x,y,z) = \frac{xy - X(y,z)y - RR(x,z) + RR(X(y,z),z)}{K(x,y,z)}$$

$$= 1 + xyz + (x^2y^2 + x^2 + y^2 + 1)z^2 + O(z^3).$$

◁

5.6.3 Models with Non-D-finite Generating Functions

How can we show that a model does not have a D-finite generating function? Three of the four criteria mentioned earlier in the chapter are useful.

Example 5.10. The set of quarter-plane lattice walks given by steps $\times$ has a non-D-finite generating function since the generating function has an infinite number of singularities. [MR09, MM16] ◁

In some cases the non-D-finiteness of $Q^{\mathscr{S}}(x,y,z)$ can be established if we can show that $Q^{\mathscr{S}}(0,0,z)$ is not D-finite. This is the generating function of excursions, and there are more general methods for excursions.

Example 5.11. The excursions of the model $\mathscr{S} = \mathrel{\text{\scriptsize K}}$ have asymptotic growth like

$$\mathrm{walk}^{\mathrel{\text{\scriptsize K}}}((0,0) \xrightarrow{n} (0,0)) \sim Cn^{\alpha}\rho^{-n}$$

with ρ algebraic, but $\alpha = -1 - \frac{\pi}{\arccos(c)}$, such that c is not a rational root of unity. Thus, α is transcendental and the excursion generating function $Q^{\mathscr{S}}(0,0,z)$ is not D-finite. Since algebraic functions are closed under algebraic substitution, this implies that the complete generating function $Q^{\mathscr{S}}(x,y,z)$ is not D-finite. This process is made algorithmic in [BRS14]. ◁

The hypertranscendency of the remaining models can be established using differential Galois theory [DHRS18]. Roughly, in some cases a natural parametrization of the kernel gives rise to equations of form Eq. (5.5).

5.7   Groups and the Co-growth Problem

Next we consider a second example where properties of generating functions elucidate the structure of a family of discrete objects. It comes from group theory, and the reader is pointed to textbooks on graduate algebra for full definitions and algebraic context as our interest is limited to the interface with generating function classes rather than the group theoretic terminology.

Given a group G with generating set $\mathscr{S}$, consider an arbitrary finite product of generators in the group: $w \in \mathscr{S}^*$. A natural question asks whether or not this product is equal to the identity in the group. To answer this question for a given group is to solve its **word problem**. There are numerous follow-up questions: What is the computational complexity of effectively computing the answer for a given word in a group with a fixed generating set? Does the complexity depend on the choice of $\mathscr{S}$? Which group theoretic properties are important to answer these questions?

This line of inquiry leads to natural enumeration questions: Given a group and a generating set, what is the number of distinct words in the set $\mathscr{S}^n$? This is the **growth problem**. A second natural question adapted to the methods we have developed so far is the co-growth problem: Determine the number of words in $\mathscr{S}^n$ that can be reduced to the identity.

We visualize a group with the aid of its **Cayley graph**. Given a group G with inverse closed generating set $\mathscr{S}$, the Cayley graph $X(G, \mathscr{S})$ is defined as the graph with vertex set $V = G$, and edge set $E = \{\{g_1, g_2\} : g_1 = g_2 s, s \in \mathscr{S}\}$. Figure 5.2 illustrates some examples. Note, we could orient and then label each edge with an element of $\mathscr{S}$: $g \xrightarrow{s} gs$.

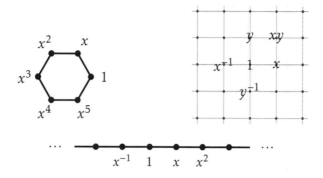

FIGURE 5.2
Cayley graphs of $G = \mathbb{Z}/6\mathbb{Z} = \langle x \mid x^6 = 1 \rangle, \mathscr{S} = \{x, x^{-1}\}$; $G = \langle x, y \mid xy - yx \rangle$; $\mathscr{S} = \{x, x^{-1}, y, y^{-1}\}$, and $G = \langle x \rangle, \mathscr{S} = \{x, x^{-1}\}$.

Structure of Groups and the Co-growth Problem

Determine analytic criteria on $C_G^{\mathscr{S}}(z)$ that correspond with group properties of G.

5.7.1 Excursions on Cayley Graphs

We restate the co-growth problem in terms of excursions on the Cayley graph. A word w in $\mathscr{S}^*$ encodes uniquely a walk on the $X(G, \mathscr{S})$ starting at 1: We follow the directed edges in order as presented in w. The number of excursions of length n from 1 to 1 on the Cayley graph $X(G, \mathscr{S})$ is denoted

$$\text{walk}_{X(G;\mathscr{S})}^{\mathscr{S}}(1 \xrightarrow{n} 1). \tag{5.13}$$

This is precisely the **co-growth sequence** of G with respect to $\mathscr{S}$. The generating function of this sequence is the **co-growth series**. Here we denote it by

$$C_G^{\mathscr{S}}(z) = \sum_{n \geq 0} \text{walk}_{X(G;\mathscr{S})}^{\mathscr{S}}(1 \xrightarrow{n} 1) z^n.$$

The walks that return to the origin form a combinatorial class that is a subclass of $\mathscr{S}^*$. Let us denote this class as the formal language $\mathscr{L}(G, \mathscr{S})$ over the alphabet $\mathscr{S}$. If a group is finite, its Cayley graph is essentially a finite automata with the vertex labelled 1 as the start and only accepting state. Anisimov also proved the converse: If the language $\mathscr{L}(G,\mathscr{S})$ is a regular language, then the group G must be finite [An171]. In this case, $C_G^{\mathscr{S}}(z)$ is a rational function. There are no known examples of groups such that $C_G^{\mathscr{S}}(z)$ is rational but not $\mathbb{N}$-rational.

The groups for which $\mathscr{L}(G,\mathscr{S})$ is specified by an algebraic grammar are also characterized. The condition is that the group G has a free subgroup of finite-index [MS83, Dun85]. These are known as the **virtually free groups**. The Cayley graphs are very tree-like. It is not known if this completely characterizes all algebraic $C_G^{\mathscr{S}}(z)$, that is, are there groups such that $C_G^{\mathscr{S}}(z)$ is algebraic, but not $\mathbb{N}$-algebraic?

Example 5.12. If

$$G = \langle x \mid x^2 = 1 \rangle \star \langle y \mid y^n = 1 \rangle \cong \mathbb{Z}/2\mathbb{Z} \star \mathbb{Z}/n\mathbb{Z}$$

and $\mathscr{S} = \{x, y\}$, then

$$C_G^{\mathscr{S}}(z) = (1 - zD)/((1 - zD)^2 - z^2),$$

where D is the unique power series solution to the equation

$$z^{n-1}(1 - zD)^{n-1} = (1 - zD - z^2)^{n-1}D$$

whose expansion begins $z^{n-1} + O(z^n)$.   ◀

Example 5.13. Let $G = \langle x \mid x^2 \rangle \star \langle y \rangle \cong \mathbb{Z}/2\mathbb{Z} \star \mathbb{Z}$ and $\mathscr{S} = \{x, y, y^{-1}\}$. Then we can compute that

$$C_G^{\mathscr{S}}(z) = \frac{1}{2} \cdot \frac{1 - 3\sqrt{1 - 8z^2}}{1 - 9z^2},$$

and $C_G^{\mathscr{S}}(z)$ has radius of convergence $\frac{1}{2\sqrt{2}}$.   ◀

5.7.2 Amenability vs. D-finiteness

Amenability is an important property of groups. There are many characterizations, although the initial definition is relative to the Tarski paradox, which is out of scope for this text. Relevant to us is the characterization of Kesten. He proved [Kes59] that for finitely generated groups, amenability can be characterized in terms of co-growth: a finitely generated group G with symmetric generating set $\mathscr{S}$ is amenable if and only if $\text{walk}_{X(G;\mathscr{S})}^{\mathscr{S}}(1 \xrightarrow{2n} 1)^{1/2n} \to |S|^2$ as $n \to \infty$.

Roughly this means that the probability that a random walk returns to 1 tends to 1 as n gets large. Amenability is an important property, and it is significant that it can be identified from subexponential growth of the co-growth sequence.

From this characterization, we see that $G = \mathbb{Z}/2\mathbb{Z} \star \mathbb{Z}$ is not amenable because the ROC is not $1/3$.

Relevant to our discussion on classification is the following result.

Theorem 5.17. *Let G be a finitely generated amenable group that is not nilpotent-by-finite and let $\mathscr{S}$ be a finite symmetric generating set for G. Then the co-growth series is not D-finite.*

This is proved using the characterization of D-finiteness in Theorem 5.7 combined with work of Kesten and Varopolous on the co-growth of amenable groups of superpolynomial growth.

Example 5.14. The lamplighter groups $L(d,H) = \mathbb{Z}^d \wr H$, where H is a finite abelian group and $d \geq 1$ are known to be amenable, and the asymptotics are incompatible with being D-finite. ◄

On the other hand, the co-growth series for the Baumslag-Solitar group $BS(k,k) = \langle x,y \mid x^k y = y x^k \rangle$ has D-finite co-growth generating series [ERJvRW14], and as it is of superpolynomial growth, we conclude that it is not amenable, except in the case $k = 1$ case, which is abelian and corresponds to simple 2D excursions:

$$BS(1,1) = \langle x,y \mid xy = yx \rangle.$$

For other k, the generating function can be written as a diagonal of an algebraic function. The D-finiteness is immediate from this characterization.

The upshot is that the generating function classification can be used to determine properties of groups.

5.8 Discussion

A series is $\mathbb{N}$-rational if it is the generating function of some regular language. The first translation from grammars to generating function is often identified as the Schützenberger methodology for regular languages. What about the other direction: Given a series, determine a regular specification whose generating function is precisely that series? Barucucci et al [BLFR01] proposed a technology for inverting rational functions to find a related S-regular grammar. Koutschan [Kou08] made this effective. Example 1.13 comes from a talk of Ira Gessel [Ges03] on this topic.

Algebraic and transcendental functions are a very classic topic of mathematics. The coefficient asymptotics in Theorem 5.2 date back to Darboux's work in the 19th century. Still, adapting the results and methods in the generating function context is an active field. Banderier and Drmota considered analytic properties of algebraic functions useful for the combinatorialist [BD15]. Notably, if a class $\mathscr{C}$ can be described by a non-ambiguous context-free grammar, then there are restrictions on the subexponential growth beyond those stated in Theorem 5.2: α in the conclusion of this theorem must come from an explicit set of values. For example, the subexponential growth cannot be $n^{1/3}$. Such a restriction implies that the Gessel walk generating function is not N-algebraic and the class cannot be generated by an unambiguous context-free grammar. We have already shown that the most straightforward encoding was not context-free, but this result frees us from looking for other possibilities – they will not exist.

Diagonals of rational functions are the largest source of transcendental D-finite functions in combinatorics. Although it is Lipshitz that we cite, Gessel and Zeilberger provided proofs later found to be incomplete. Chen did eventually fill in the gaps in their argument [WC13].

A potential distinct source was thought to be pattern-avoiding permutations. However, Garrabrant and Pak proved that there do exist finite (though large) pattern sets whose avoidance class is not D-finite. Many smaller avoidance classes are in bijection with lattice walks, such as Baxter permutations [CMMR17].

It is conjectured that the Catalan generating function cannot be written as a diagonal of an N-rational function [GP], although we have seen two different expressions as diagonals of differences of N-rational functions.

To see detailed examples of how to show generating function is not D-finite, there is an interesting series of articles on restricted polygon shapes on lattices [Rec06a, Rec06b, Rec09]. Theorem 5.5 is used and is proved in [BMR03]. Another approach that we could take to study combinatorial classes with D-finite functions is to make combinatorial sense of differentiation. On the level of series, this seems like it should be straightforward: since $[z^n]F'(z) = (n+1)f_{n+1}$. The change of size poses a problem– the operator that is more straightforward to interpret is $\Theta := z\frac{d}{dz}$. These are not equivalent! Not every linear differential equation can be written in terms of Θ. The interpretation views Θ as a pointing operator which has an easy interpretation for labelled objects. [Gre83, BFKV11, BFKV07]. Pointing can be used to define set and cycle constructions, particularly when paired with an "unpointing" operator: A pointed cycle is a sequence and a pointed set is an object and smaller set. The pointing naturally breaks symmetry and makes things easier to count.

The study of properties of series solutions of differential equations is also a very classic topic of mathematics. Even standard references such as [Was65] contain results very applicable to combinatorial enumeration problems.

Theorem 5.16 is proved by using functional equations in [BM02], and then combinatorially in a weaker case in [BMP00].

Theorem 5.7 appears in [BM18], but appears earlier in the literature in German [Rem95].

Guttmann and coauthors has developed a very useful suite of analyses to test D-finiteness of combinatorial generating functions [Gut00].

The kernel in the equations of the two-dimensional quarter-plane lattice walks can be parametrized using elliptic curves, and this helps to explain some of the taxonomic structure from an analytic perspective. The survey of Bostan [BOS] is very useful to understand how to use computer algebra in this study [Bos]. Beyond this case, there are some variants. Step sets with negative steps of length 2 have been thoroughly considered [BBMM]. Walks in three dimensions, and excursions in arbitrary dimensions have also been considered. Restricting to nonconvex cones reveals a similar taxonomy [BM16, RT].

Garrabant and Pak [GP17, Conjecture 13] first conjectured Theorem 5.17 and proved the the conjecture held for numerous classes, including virtually solvable groups of exponential growth with finite Prüfer rank; amenable linear groups of superpolynomial growth: groups of weakly exponential growth $A\exp(n^\alpha) < \gamma_{G,S}(n) < B\exp(n^\beta)$ with $A, B > 0$, and $0 < \alpha, \beta < 1$, where $\gamma_{G,S}(n)$ is the number of distinct elements of G that can be expressed as a product of n elements of S.

5.9 Problems

Exercise 5.1. Prove that an algebraic power series is also D-finite. Prove that $e^z = \sum_{n \geq 0} z^n/n!$ is not an algebraic series but that it is D-finite to conclude that algebraic series are a proper subset of the D-finite series. ❑

Exercise 5.2. Show that if $F(z) = \sum f_n z^n$ is D-finite to order r, and that each polynomial in Eq. (5.2) is of order at most d, and there at least one polynomial of degree d, then the coefficient sequence is P-recursive to order at most $r + d$, and the degree of the polynomials is bounded by r. Find the inverse statement.

Reference: [BCG⁺17, Theorem 14.1]

❑

Exercise 5.3. The shuffle of two languages is the union of the pairwise shuffle of words. Show that the **shuffle** two context-free languages is not necessarily context-free, but its generating function is still D-finite. Use this to make a direct argument that the generating function for simple excursions confined to the quarter-plane has a D-finite generating function.

Reference: [FGT92]

❑

Exercise 5.4. Can a function with a natural boundary be differentiably algebraic? ❑

Exercise 5.5. Determine the k for which $\sum_n \binom{2n}{n}^k z^n$ is transcendental D-finite.

Reference: [Sta80]

❑

Exercise 5.6. Show that there are 79 non-trivial small-step models of walks in the quarter-plane. Try to estimate the number in three dimensions.

Reference: [BMM10, BBMKM16]

❑

Exercise 5.7. Show that $\sum_{n\geq 0} \pi(n) z^n$ is not D-finite if $\pi(n) = \#$ of primes less than n. It is open to determine if it is hypertranscendental or not.

Reference: [FGS06]

❑

Exercise 5.8. Suppose that $\sum_{n\geq 0} \frac{a_n}{n!} z^n$ is D-algebraic but not D-finite. Does this imply that $\sum_{n\geq 0} a_n z^n$ is hypertranscendental?

Reference: [Pak18, Open Problem 2.4]

❑

Exercise 5.9. Is $\frac{1+x}{1+z-2z^2-3z^3}$ N-rational? Either prove that it is not or express it in a manner that makes N-rationality obvious.

Reference: [Kou08][Ges03]

❑

Exercise 5.10. In his fundamental 1980 paper, Stanley conjectured the transcendency of the series

$$\sum_{n\geq 0} \binom{2n}{n}^r z^n$$

for integer r. Prove what you can. (Hint: start with $r = 1, r = 2,$ even r)

Reference: [Sta80][Fla87][WS89]

❑

Exercise 5.11. Prove that

$$F(z) = \sum z^{k^n}$$

is hypertranscendental for any $k \in \mathbb{N}$ greater than 1. ❑

Exercise 5.12. Is $\sum \tau(n) z^n$ where $\tau(n)$ is the number of divisors of n D-algebraic? ❑

Exercise 5.13. Characterize the bivariate rational functions whose diagonals are N-algebraic. ❑

Exercise 5.14. Consider the following language over the alphabet $\Sigma = \{a, b\}$: $\mathcal{L} = \{a^n b w_1 a^n w_2 \mid w, w_2 \in \Sigma^*, n \in \mathbb{N}\}$. This language is context-free, but for every grammar that defines it there exist words with multiple derivation trees. That is, the combinatorial class is not given by an algebraic specification. To prove such a tricky non-existence result, first show that the generating function $L(z)$ of the language is

$$L(z) = \sum_{n \geq 1} \frac{z^{2n}}{1 - 2z + z^{n+1}}.$$

Show that this generating function is transcendent since it has poles that accumulate near $1/2$. Is it D-finite? Is it D-algebraic?

Reference: [Fla87]

☐

6

Singularities of Multivariable Rational Functions

CONTENTS

6.1   Visualizing Domains of Convergence 140
    6.1.1   The Univariate Case 140
    6.1.2   The Multivariable Case 141
6.2   The Exponential Growth 144
6.3   The Height Function 146
6.4   Visualizing Critical Points 148
6.5   Examples 149
    6.5.1   Delannoy Numbers 150
    6.5.2   Balanced Words 151
6.6   Discussion 153
6.7   Problems 153

The location of the boundary of convergence of a formal power series gives information about the growth of its coefficients. This is as true in the multivariate case as it is in the univariate case. Given a multivariable rational function with a series expansion we want to understand the growth of the coefficients along a fixed ray. To do this we first visualize the domain of convergence. This is not an easy task in multidimensional complex space! Since a power series converges absolutely on a domain, we can start by looking at the real points for which it converges.

Once we have a picture in mind, we can bound the coefficient growth for a diagonal of the series using elementary analysis and convergence arguments. We use some powerful results to give straightforward conditions for when these bounds are actually equalities. This process determines a set of critical points that are the analogues of the dominant singularities in univariate case. These points drive the dominant terms in the asymptotic expansions.

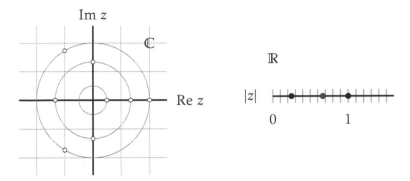

FIGURE 6.1
(left) The eight complex singular points of $\frac{1}{(1-z^3)(1-4z)^2(1-5z^4)}$. *(right)* The image of the complex plane under the map $z \mapsto |z|$. The image of the singularities break the line into four segments, each corresponding to an annulus where a Laurent series is defined.

6.1 Visualizing Domains of Convergence

Let us fix now the dimension d, and consider a function $F(\mathbf{z}) : \mathbb{C}^d \to \mathbb{C}^d$ equipped with a series development $\sum_{\mathbf{n}} f(\mathbf{n})\mathbf{z}^{\mathbf{n}}$ around the origin.

6.1.1 The Univariate Case

In the univariate case, the dominant singularities of $F(z)$ have smallest modulus amongst all its singularities. This modulus is the radius of convergence and gives the exponential growth of the coefficient sequence (f_n):

$$\text{ROC}\left(\sum f_n z^n\right) = R \quad \Longrightarrow \quad \limsup_{n \to \infty} f_n^{1/n} = R^{-1}.$$

In Figure 6.1 we map $\mathbb{C}$ to $\mathbb{R}_{\geq 0}$ using $z \mapsto |z|$. Singularities on the same circle are mapped to the same point. The image on the right is a schematic for the different domains of convergence: Note, they are a collection of disjoint connected regions. In this figure, points may represent several singularities in the pre-image and also points that are not singular. We will determine a similar figure for the functions of several variables.

6.1.2 The Multivariable Case

A function can have multiple series developments, each valid in a given (unique) domain that can be visualized in $\mathbb{R}^d$. In combinatorics, we are interested in the domain that includes the origin.

First we should be precise about what it means for a multivariate series to converge. If we use multi-index notation, $\sum_{\mathbf{n} \in \mathbb{N}^d} f(\mathbf{n}) \mathbf{z}^{\mathbf{n}}$, we interpret this as nested sums:

$$\sum_{n_1 \geq 0} \left(\cdots \left(\sum_{n_d \geq 0} f(\mathbf{n}) z_1^{n_1} \cdots z_d^{n_d} \right) \cdots \right).$$

The value of the summation can depend on the order of summation. The combinatorial context is a slightly less general situation, as our series are either grounded by a variable marking size, or the series that have all positive integer terms, but it is important to be aware of this subtlety. That said, the terms of an **absolutely convergent series** can be reordered arbitrarily without changing the value of the sum (or the convergence of the sum). The **domain of convergence** of a power series, denoted here with $\mathcal{D}$, is an open, connected set formed by the interior of the set of points at which the series converges absolutely. Like the univariate case, the domain of convergence is **multicircular**. The relevant vocabulary is as follows. The **polydisk** of a point z is the domain

$$D(z) := \{z' \in \mathbb{C}^d \mid |z_i'| \leq |z_i|, 1 \leq i \leq d\}.$$

The **torus** associated to a point is

$$T(z) := \{z' \in \mathbb{C}^d \mid |z_i'| = |z_i|, 1 \leq i \leq d\}.$$

A domain of convergence is **multicircular**: if a point $\mathbf{z} = (z_1, \ldots, z_d)$ lies in the domain, then the domain also contains the set of points $T(z)$, by absolute convergence.

The set of **singularities** of a rational function $F(\mathbf{z}) = G(\mathbf{z})/H(\mathbf{z})$ is precisely the set

$$\mathcal{V} := \{\mathbf{z} \in \mathbb{C}^d \mid H(\mathbf{z}) = 0\}.$$

This is a variety, **the singular variety of** F. We are interested in singularities that are "closest" to the origin. We define closest componentwise giving the following definition. Our working definition of minimal point is the following. The **minimal points** of a multivariate series are the points from $\mathcal{V}$ on the boundary $\partial \mathcal{D}$ of domain of

convergence under the constraint that no coordinate is zero. Said an-
other way, a point $\mathbf{z}$ in the variety $\mathcal{V}$ is minimal if there is no point
$\mathbf{z'} \in \mathcal{V}$ such that $\left|z'_j\right| < \left|z_j\right|$ for *all* j from 1 to d. A minimal point is
strictly minimal if it is the only minimal point on its torus, that is
$\mathcal{V} \cap T(\mathbf{z}) = \{\mathbf{z}\}$.

These definitions naturally extend the univariate case: If a function is not entire, it has a singularity on its boundary of convergence.
The singularities on the circle of convergence are the minimal points.
If there is only one such singularity, then it is a strictly minimal point
and will control the dominant asymptotics alone.

Recall a series is said to be **combinatorial** if the coefficients are all
nonnegative integers. In the $d = 1$ case this forced the existence of a
dominant singularity on the positive real line (Pringsheim's lemma).
Given the mechanics of the proof, it might not be surprising that this
generalizes to higher dimensions.

Lemma 6.1. *Suppose that $F(\mathbf{z}) \in \mathbb{N}[[\mathbf{z}]]$ is a combinatorial series. The
point $\rho \in \mathbb{C}^d \in \mathcal{V}$ is a minimal point of $F(\mathbf{z})$ if and only if the point
$(|\rho_1|, \ldots, |\rho_d|) \in \mathbb{R}^d$ is a minimal point.*

This result is proved as Theorem 3.16 in [PW13] and was slightly
generalized by Melczer and Salvy. Since we are often in the combinatorial case, this information is very useful to find minimal points.

Example 6.1. Let $F(x,y) = \frac{1}{1-x-y}$. Let us revisit the definitions.

Singular Variety The singular variety $\mathcal{V}$ is the set of points $(x,y) \in \mathbb{C}^2$
annihilating $1 - x - y$:

$$\mathcal{V} = \{(x, 1-x) \in \mathbb{C}^2 \mid x \in \mathbb{C}\}.$$

Minimal Points Figure 6.2 illustrates the image of $\mathbb{C}^2$ under $(x,y) \mapsto$
$(|x|, |y|)$. Consider the boundary on the bottom left edge of the region marked $\mathcal{V}$. These are points (x,y) such that there is no point
(x',y') in $\mathcal{V}$ with $|x'| < |x|$ and $|y'| < |y|$: There is no point in the
image of $\mathcal{V}$ to the left, and underneath these points. Thus, the
pre-image of this line in $\mathcal{V}$ comprises the set of minimal points.
Points in the pre-image satisfy $|1 - x| = 1 - |x|$, which only holds
when $x \in \mathbb{R}_{\geq 0}$. The minimal points are:

$$\{(x, 1-x) \mid x \in \mathbb{R}_{\geq 0}\}.$$

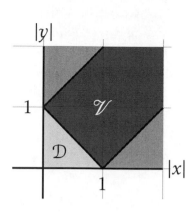

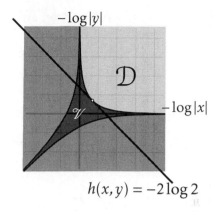

FIGURE 6.2

The singular variety $\mathcal{V}$ and domain of convergence $\mathcal{D}$ of Example 6.1. The light grey areas correspond to domains of convergence for other series expansions of $\frac{1}{1-x-y}$. *(left)* The image of $\mathbb{C}^2$ under $(x, y) \mapsto (|x|, |y|)$. *(right)* The image of $\mathbb{C}^2$ under the -relog map $(x, y) \mapsto (-\log|x|, -\log|y|)$. The line $h(x, y) = -2\log 2$ is tangent to the domain.

Strictly Minimal Points If two minimal points $(x, 1-x)$ and $(x', 1-x')$ are on the same torus, then $|x| = |x'| \implies x = x'$ since they are both positive real numbers. Thus, each minimal point is the only one on its torus, and so all of the minimal points are strictly minimal.

Domain of Convergence We are interested in the Taylor series expansion obtained by expanding the geometric series:

$$F(x, y) = \sum_{\ell \geq 0} \sum_{k \geq 0} \binom{\ell + k}{k} x^k y^\ell.$$

We can determine the domain of convergence. We consider the variables one at a time. First, let us consider the interior sum for any fixed positive integer ℓ. The summation is an instance of the generalized binomial theorem:

$$\sum_{k \geq 0} \binom{\ell + k}{k} x^k = (1 - x)^{-\ell+1}.$$

It is clear that $|x| < 1$ when this converges. Because the coefficients are all positive, the order does not matter, and we can similarly conclude that $|y| < 1$. This gives us a first understanding of our series.

Now, if the series is convergent at (x, y) it is convergent at $(|x|, |y|)$, so we consider $\frac{1}{1-|x|-|y|}$. This geometric series converges if and only if $|x| + |y| < 1$. The domain of convergence is thus

$$\mathcal{D} = \{(x, y) \in \mathbb{C}^2 \mid |x| + |y| < 1\}.$$

The boundary of this domain is $\partial \mathcal{D} = \{(x, y) \in \mathbb{C}^2 : |x| + |y| = 1\}$, and we confirm that this is precisely the constraint of the minimal points.

Figure 6.2 illustrates different connected regions under the map $(x, y) \mapsto (|x|, |y|)$, notably the three distinct regions of convergence, their boundaries. The image of $\mathcal{V}$ under this map is the region in dark grey. ◄

6.2 The Exponential Growth

As in the univariate case, minimal points determine bounds on the exponential growth. Suppose that $\mathbf{z}$ is a point inside the boundary of convergence of the series $\sum_{\mathbf{n}} f(\mathbf{n}) \mathbf{z}^{\mathbf{n}}$. Since convergent power series are absolutely convergent, the series

$$\sum_{\mathbf{n}} f(\mathbf{n}) |z_1|^{n_1} |z_2|^{n_2} \ldots |z_d|^{n_d} \text{ is convergent.}$$

If a series is convergent, then so are its subseries, consequently

$$\sum_{n \geq 0} f(n, n, \ldots, n) |z_1|^n |z_2|^n \ldots |z_d|^n \text{ is also convergent.}$$

As this is equal to $\sum f(n, n, \ldots, n) |z_1 z_2 \ldots z_d|^n$, we have that the point $|z_1 z_2 \ldots z_d|$ is within the radius of convergence of $\Delta F(\mathbf{z})$. We bound the exponential growth by examining the closure of the domain of convergence:

$$\limsup_{n \to \infty} |f(n, n, \ldots, n)|^{1/n} \leq |z_1 z_2 \ldots z_d|^{-1} \quad \text{with } (z_1, \ldots, z_d) \in \overline{\mathcal{D}}$$

We find

$$\limsup_{n\to\infty} |f(n,n,\ldots,n)|^{1/n} \le \inf_{(z_1,\ldots,z_d)\in\partial D} |z_1 z_2 \ldots z_d|^{-1}.$$

This is a bound – it might be that the diagonal is empty. We give conditions under which this is an equality in the next section. This is a substantial result. Pemantle and Wilson add the constraint that if in a neighbourhood of the ray the diagonal is defined, then this bound is indeed equality.

In our combinatorial cases, this condition is frequently met, and the supremum is reached on a point of the variety $\mathcal{V}$:

$$\rho = \sup_{z\in\overline{D}\cap\mathcal{V}} |z_1 \ldots z_d|. \tag{6.1}$$

Example 6.2. We return to our example: $\frac{1}{1-x-y}$ with $f_{k,\ell} = \binom{k+\ell}{k}$. Then

$$\limsup_{n\to\infty} f_{n,n}^{1/n} = \inf_{(x,y)\in\partial D} |xy|^{-1} = \inf_{x\in\mathbb{R}}(x(1-x))^{-1} = 4.$$

Because the series is combinatorial, to find the exponential growth we can restrict our attention to the positive real solutions. The value we find is consistent with the limit $\lim_{n\to\infty} \binom{2n}{n}^{1/n} = 4$ obtainable by Stirling's approximation. Similarly,

$$\limsup_{n\to\infty} f(rn,sn)^{1/n} = \inf_{(x,y)\in\partial D} |x^r y^s|^{-1} = \inf_{x\in\mathbb{R}}(x^r(1-x)^s)^{-1}.$$

This is minimized at $x = \frac{r}{r+s}$ from which we deduce an expression for the exponential growth:

$$\left(\left(\frac{r}{r+s}\right)^r \left(\frac{s}{r+s}\right)^s\right)^{-1}.$$

◄

Now we have a strategy: First find the minimum points by considering singular points on the domain of convergence. When the rational function is combinatorial, and we are interested in the exponential growth, we can consider only the real solutions. We then minimize $|z_1^{r_1}\ldots z_d^{r_d}|^{-1}$ among these points.

In the next section, we will need all of the minimizing solutions, not just the real ones, to determine the subexponential growth.

6.3   The Height Function

The minimization we require involves a nonlinear function. It is easier to minimize a linear function. Define a function $h : \mathcal{V}^* \to \mathbb{R}$:

$$h : (z_1, \ldots, z_d) \mapsto -\log |z_1| - \cdots - \log |z_d|. \tag{6.2}$$

We call this function the **height function**. It a real valued function defined in the domain $\mathcal{V} \setminus \{\mathbf{z} : z_1 \ldots z_d \neq 0\}$. The map h is smooth, and hence it is minimized at its critical points, determined in the manner of classic calculus. In order to work with this function and visualize how it interacts with the domain of convergence, we map $\mathbb{C}^d$ to $\mathbb{R}^d$ under the **-relog** map:

$$\mathrm{relog} : \mathbb{C}^d \to \mathbb{R}^d$$

$$-\mathrm{relog} : \mathbf{z} \mapsto (-\log |z_1|, \ldots, -\log |z_d|).$$

This map helps us understand domains of convergence. If $F(z)$ is a Laurent series, then the domain of convergence of F has the form $\mathcal{D} = \mathrm{relog}(B)$ for some convex open set B [PW13, Thm. 7.2.2].

In this coordinate system, $h(\mathbf{z}) = c$ defines a hyperplane as we see in Figure 6.2 (right). To minimize $\left| z_1^{r_1} \ldots z_d^{r_d} \right|^{-1}$ on $\partial \mathcal{D}$, we first find the minimum c so that the hyperplane $h(\mathbf{z}) = c$ touches the image of $\mathcal{V}$ in the image of the relog map. Each contact can potentially represent multiple points in the pre-image of $-\mathrm{relog}$ (in $\mathbb{C}^d$).

If $H(\mathbf{z})$ is a Laurent polynomial then it turns out that

$$\mathbb{R}^n \setminus \{-\mathrm{relog}(\mathbf{z}) \mid \mathbf{z} \in \mathcal{V}\}$$

is a collection of convex regions. Each real convex set is in bijection with a a Laurent series expansion of the rational function $\frac{1}{H(\mathbf{z})}$. If it has a power series expansion, then under this bijection it corresponds to the component containing $(n, n, \ldots, n)$ for n sufficiently large.

In Figure 6.2 we see three regions, each corresponding to a series development. The function $\frac{1}{1-x-y}$ has a Taylor series expansion, and it corresponds to the region labelled $\mathcal{D}$, as we have indicated.

In this text, we are only interested in Taylor expansions, and hence going forward we will only draw the component relevant to our series extraction (more precisely, we draw its boundary), but it is useful to know that the process described here does work for other series expansions.

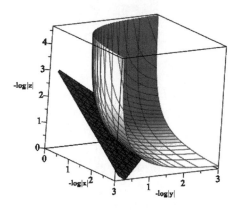

FIGURE 6.3
The image of the boundary of convergence for the Taylor series of $F(x,y,z) = \frac{1}{1-x-y-z}$ under the -relog map (white) and the hyperplane $h(x,y,z) = 3\log 3$ (dark grey).

The Critical Point Equations

$$H(\mathbf{z}) = 0, \quad r_1^{-1} z_1 \frac{\partial H(\mathbf{z})}{\partial z_1} = r_k^{-1} z_k \frac{\partial H(\mathbf{z})}{\partial z_k}, \quad k = 2,\ldots,d$$

The critical points are solutions to the following system, called the **critical point equations**:

$$H(\mathbf{z}) = 0, \quad r_1^{-1} z_1 \frac{\partial H(\mathbf{z})}{\partial z_1} = r_k^{-1} z_k \frac{\partial H(\mathbf{z})}{\partial z_k}, \quad k = 2,\ldots,d. \qquad (6.3)$$

In what follows, we refer to the solutions as **critical points**, although they are defined geometrically. The geometry of the variety $\mathscr{V} = \{(z_1,\ldots,z_d) : H(z_1,\ldots,z_d) = 0\}$ at critical points will decide the sub-exponential growth.

Here is the formal statement.

Proposition 6.2. *Let H be an irreducible polynomial with associated variety $\mathscr{V}$. Then $\rho \in \mathscr{V}$ is a critical point for $\mathbf{r} \in \mathbb{N}^d$ if and only if $\mathbf{r}$ can be* *written as a linear combination of the vector*

$$\left(z_1 \frac{\partial H(\mathbf{z})}{\partial z_1}, z_2 \frac{\partial H(\mathbf{z})}{\partial z_2}, \ldots, z_d \frac{\partial H(\mathbf{z})}{\partial z_d} \right).$$

We generalize this later once we have a bit more sense of the geometry (See Proposition 8.1). A critical point is said to be **strictly minimal** if it is on the boundary of convergence of the series. It is additionally qualified to be **finitely minimal** if there are only a finite number of strictly minimal points points. A critical point is **isolated** if there exists a neighbourhood of $\mathbb{C}^d$, where it is the only critical point. In the generating function we consider in this text the asymptotics are driven by isolated finitely minimal critical points. They are isolated singular points of a rational expression, and there is no further cancellation from the numerator. This is the easiest generic case to study. Later we will also identify **contributing points**. These are the points that actually contribute to the dominant term in the asymptotic expansion of the coefficients.

6.4 Visualizing Critical Points

What is the height function? We can best understand it by looking at the regions of convergence under that transformation.

Example 6.3. The result of the height function matches what we have found already in our examples. For example, $1 - x - y = 0 \implies x + y = 1$ and

$$x \frac{\partial H(x,y)}{\partial x} = y \frac{\partial H(x,y)}{\partial y} \implies x = y \implies x = y = 1/2.$$

The critical point is at $(1/2, 1/2)$. We can see this in the logarithmic domain of convergence. The line $h(x,y) = 2\log(2)$ touches $\partial \mathcal{D}$ at $(-\log(2), -\log(2))$.   ◀

Figure 6.3 illustrates the geometry in the case of $F(x,y,z) = \frac{1}{1-x-y-z}$. The height function h for the ray $\mathbf{r} = (1,1,1)$ defines a hyperplane $h(x,y,z) = 3\log 3$ that touches the boundary of convergence at a unique point $-\log 3 \cdot (1,1,1)$.

6.5   Examples

So far we have seen examples when H is a single polynomial. The more general case is a little more subtle, and we consider it in Chapter 8. If H factors nontrivially into square-free factors, $H = H_1 \dots H_k$ and ρ is a root of one of the factors, and it is not a root of the other factors: $H_k(\rho) = 0 \implies H_\ell(\rho) \neq 0$ for $i \neq j$, then a similar process can be invoked. If H is not square-free, then an analogous result is true once we remove the repeat factors. Take note! Critical points are **potential locations of minimizers** of $|z_1 \dots z_d|^{-1}$. In the most straightforward cases, it suffices to compare the values of this product and select the critical point that is the global minimizer. Indeed, the best case if if there are a finite number or even a single unique minimal critical point. In the general case identifying the critical point, and deciding which ones contribute to the asymptotics is a difficult problem.

Estimating Exponential Growth

Suppose that $F(\mathbf{z}) = \frac{G(\mathbf{z})}{H(\mathbf{z})}$ (H an irreducible polynomial) with series expansion $\sum_\mathbf{n} f(n)\mathbf{z^n}$ with domain of convergence $\mathcal{D}$. Let $\mathbf{r} \in \mathbb{N}$ be such that $f(\mathbf{n})$ is non-zero for $\mathbf{n}$ is a small neighbourhood of $n\mathbf{r}$.

- Solve the critical point equations in the domain of $\partial\mathcal{D}$;

- If the critical points are minimal, determine the points $\mathbf{z}$ for which $\left|z_1^{r_1} \dots z_d^{r_d}\right|^{-1}$ is globally minimized amongst the critical points;

- Conclude $\limsup_{n \to \infty} f_{n\mathbf{r}}^{1/n} = \left|z_1^{r_1} \dots z_d^{r_d}\right|^{-1}$.

6.5.1   Delannoy Numbers

We return to the case of Delannoy numbers of Example 3.5 which was equivalent to studying the following diagonal.

$$\Delta^{(r,s)} \frac{1}{1 - x - y - xy} \implies H(x,y) = 1 - x - y - xy.$$

We use Gröbner bases to solve the critical point equations. We find the basis of [H, s*x*diff(H, x)-r*y*diff(H,y)] with respect to

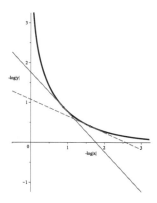

FIGURE 6.4
The image of the boundary of convergence for the Delannoy generating function. The solid line is the curve $(-\log|x|, -\log|y|) : (x,y) \in \partial D$. The grey line finds the critical point for $(r,s) = (1,1)$ and the dashed line finds the critical point for $(r,s) = (5,12)$.

the PLEX ordering. This returns $[-s - s + sy^2 + 2ry, s - sy + r + x]$. We can solve for y, in the first component, and then for x using this value for y. We get a critical point at

$$\left(\frac{\sqrt{r^2 + s^2} - s}{r}, \frac{\sqrt{r^2 + s^2} - r}{s} \right).$$

Figure 6.4 illustrates the image of the relative position of critical point for $(r,s) = (1,1)$ and $(r,s) = (5,12)$. Thus, the exponential growth factor of the sequence $d_{nr,ns}$ is:

$$\left(\frac{\sqrt{r^2 + s^2} + s}{r} \right)^r \left(\frac{\sqrt{r^2 + s^2} - r}{s} \right)^s,$$

which is $3 + 2\sqrt{2}$ for the central Delannoy numbers ($r = s = 1$).

6.5.2   Balanced Words

Let $\mathscr{L}$ be the combinatorial class of binary words over $\{0, 1\}$ such that no word has a run of 1s of length 3 or longer. Such words avoid contiguous 1s. For example, 0101110 is not a word in $\mathscr{L}$. The following regular expression determines $\mathscr{L}$:

$$\mathscr{L} \equiv (\epsilon + 1 + 11) \cdot (0 \cdot (\epsilon + 1 + 11))^*. \tag{6.4}$$

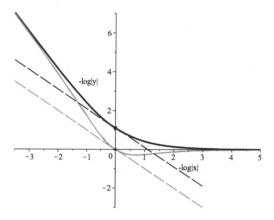

FIGURE 6.5
The image of the boundary of convergence of $L(x,y)$. The black solid line is the curve $(-\log|x|, -\log|y|) : (x,y) \in \partial\mathcal{D}$, and the grey solid line the image of a second curve in the variety. The two height functions tangent to these curves are indicated.

The bivariate generating function where x marks 1s and y marks 0 is

$$L(x,y) = \sum_{j,k} a(j,k)x^j y^k = \frac{1+x+x^2}{1-y(1+x+x^2)}.$$

Here $a(j,k) = $ # words with j 1s and k 0s, Let $\mathcal{L}_=$ be the balanced sublanguage of $\mathcal{L}$ where additionally #1s $=$ #0s:

$$\mathcal{L}_= = \{\epsilon, 01, 10, 0011, 0101, 1010, 1100, \ldots\}.$$

This is provably not a regular language, but the generating function is expressible as a diagonal: $L_=(y) := \sum a(n,n)y^n = \Delta L(x,y)$. To determine the exponential growth, we first find the Gröbner basis of the critical point equations: $[x^2-1, 3y+x-2]$. There are two solutions to this set of equations $(1, 1/3)$, and $(-1, 1)$. Recall if a point (x,y) is on the boundary of convergence of a power series, then so is $(|x|, |y|)$. In this case, we see that as $(1,1)$ is not on the boundary of convergence, it is not a critical point, but $(1, 1/3)$ is. Consequently, $\limsup_{n\to\infty} a(n,n)^{1/n} = 3$. In Figure 6.5, we see the image of the two solutions to the critical point equation. The top on is on the curve that defines the domain of convergence, and the bottom is not.

In the exercises we consider other classes of words that avoid runs.

6.6 Discussion

Determining the critical points is perhaps one of the more subtle aspects of analytic combinatorics in several variables. In truth, the cases we consider here represent the easiest cases: Once we leave the combinatorial case, stranger behaviour can happen.

The minimizer of the upper bound of $|z_1 \dots z_d|^{-1}$ on $\partial \mathcal{D}$ is not always a critical point. It is possible in the non-combinatorial case there are no solutions to the critical point equations. Melczer [Mel17, Ex. 64] gives the example $F(x,y) = \frac{1}{2+y-x(1+y)^2}$, which has no solutions to the critical point equations, but still some points on $\mathcal{V}$ become local minimizers after application of the $-$relog map. The reader is directed toward [PW13] and the references therein to consider more advanced situations such as this one.

The process of determining which critical points contribute to the dominant asymptotics amounts to comparing the moduli of the components. Melczer and Salvy [MS16] describe numerical methods to perform this filtration in the combinatorial case. Remarkably, they are able to tightly bound the complexity of the computations and the data representations.

6.7 Problems

Exercise 6.1. Find the two series expansions of $\frac{1}{1-x-y}$ whose domains of convergence are valid in the other two regions. (Reference [Mel17, Ex. 28]) □

Exercise 6.2. Describe the domain of convergence of $F(\mathbf{z}) = \frac{1}{1-z_1-z_2-\cdots-z_d}$. Use this to estimate the growth of the multinomial $\binom{dn}{n,\dots,n}$ as n tends to infinity. □

Exercise 6.3. Determine the domain of convergence and the critical points for $F(x,y,z) = \frac{1}{1-xyz(x+1/x+y+1/y)}$. Estimate the growth of $[z^n]\Delta F(x,y,z)$ as n tends to infinity. □

Exercise 6.4. Consider the class $\mathcal{F}$ of binary words over $\{a,b\}$ such that no word has a run of k a's. This is given by the grammar $\mathcal{F} = \mathcal{A}_k(b \times \mathcal{A}_k)^*$ $\mathcal{A}_k = \epsilon + \cup_{j=1}^k a^k$. If $A_k(x) = 1 + \cdots + x^{k-1} = \frac{1-x^k}{1-x}$, we conclude

$F(x,y) = \frac{A_k(x)}{1-yA_k(x)}$. For any fixed k, we can determine the exponential growth μ of the number of balanced words in $\mathcal{F}$ – the number of words of length $2n$ that have the same number of a's and b's.

Prove that for any $k > 2$ the sublanguage of $\mathcal{F}$ with $\#a = \#b$ is not S-regular, i.e., it cannot be specified directly by an iterative grammar. Under which conditions is $\Delta F(x,y)$ algebraic? For $k = 5$, determine the exponential growth for the subset of balanced words satisfying $\#a = \#b$. ❏

Exercise 6.5. Show that if $F(z)$ is combinatorial then $\rho \in \mathcal{V}$ is a minimal point if and only if $(|\rho_1|,\ldots,|\rho_d|) \in \mathcal{V}$ and the line segment $\{(t|\rho_1|,\ldots,t|\rho_d|) \in \mathcal{V} \mid 0 < t < 1\}$ from the origin to $(|\rho_1|,\ldots,|\rho_d|)$ does not contain an element of $\mathcal{V}$.

Reference: [Mel17, Proposition 62]

❏

7

Integration and Multivariable Coefficient Asymptotics

CONTENTS

7.1   A Typical Problem 156
7.2   Warm-up: Stirling's Approximation 157
7.3   Fourier-Laplace Integrals 159
7.4   Easy Inventory Problems 160
7.5   Generalizing the Strategy to Higher Dimensions 162
       7.5.1   Multivariate Cauchy Integral Formula 162
       7.5.2   A Formula for Fourier-Laplace Integrals 162
       7.5.3   How Not to Transform This Integral 163
7.6   Example: Simple Walks 164
       7.6.1   Exponential Growth 164
       7.6.2   Estimating Cauchy Integrals 165
7.7   A More General Strategy 167
7.8   Discussion 172
7.9   Problems 174

In the previous chapter, we related the domain of convergence of a multivariable series expansion of a rational function to a simplified visualization of its singular variety. From this we were able to determine bounds on the exponential growth, essentially confirming how the first principle of coefficient asymptotics is valid also for higher dimensional objects.

Does the second principle similarly generalize? If so, can we characterize what is meant by the nature of the singularity in higher dimension? The answer is yes: from a high-level point of view we replicate the proof of Theorem 4.11. The coefficient can be written as a Cauchy integral over a contour centered around the origin. By considering the same integrand but with a contour that contains the dominant singularities we can estimate the coefficient up to a known error by a sum of residue integrals. The proof the multivariate case requires significant attention to detail and requires some strong simplifying assumptions to be accessible with the background we have developed so far.

To introduce ourselves to the underlying manipulations, we consider a family of rational functions, very common in probability theory, that have smooth minimal critical points. The singular variety is smooth around the point. The asymptotic growth of the coefficients along a particular ray are are driven by these points.

7.1   A Typical Problem

In this chapter, we on to solving the following: Given $F(z_1,\ldots,z_d) = \frac{G(z_1,\ldots,z_d)}{H(z_1,\ldots,z_d)}$, a rational function with Laurent series expansion $\sum f(n_1,\ldots,n_d) z_1^{n_1} \ldots z_d^{n_d}$, and a fixed $\mathbf{r} \in \mathbb{N}_+{}^d$, determine the asymptotic growth of $f(n r_1,\ldots,n r_d)$ as n tends to infinity.

In this chapter, we will require that G and H are holomorphic on a non-empty domain $\mathcal{D} \subset \mathbb{C}^d$ and that F is not entire. Indeed, we will concentrate on the case that G and H are relatively prime polynomials and H is nontrivial and irreducible. Many of the results are true in a more general context.

The critical points we determined in the previous chapter by solving a system of polynomial equations each contribute to the asymptotic growth. They are points on the singular variety $\mathcal{V} = \{\mathbf{z} : H(\mathbf{z}) = 0\}$ on the boundary of convergence, which can be determined by solving a system of polynomial equations. Consider a minimal critical point ρ. In this chapter, we require that $\mathcal{V}$ is **smooth** at its critical points. That is,

$$\nabla H(\mathbf{z}) \neq 0 \text{ for all } z \text{ in a neighbourhood of } \rho.$$

Modulo some non-degeneracy conditions, there exist computable complex constants C_k such that for all positive integer N,

$$f(n r_1,\ldots,n r_d) = \left(\rho_1^{r_1} \cdots \rho_d^{r_d}\right)^{-n} \left[\sum_{k=0}^{N-1} C_k n^{-(d-1)/2-k} + O(n^{-(d-1)/2-N})\right].$$

We have seen how to determine the exponential growth factor in the previous chapter, and in this chapter we address how to compute the C_k which also determine the critical exponent.

We start with a classic problem with a straightforward structure: $H(\mathbf{z}) = 1 - z_d P(z_1,\ldots,z_{d-1})$ for some polynomial P. This permits us to

Smooth Point Asymptotics for Strictly Minimal Critical Points

- Input: A single dominant critical point ρ for $F(z_1,\ldots,z_d)$ in the direction $\mathbf{r}$

    - Minimal: $\mathscr{V} \cap T(\rho) = \rho$.
    - Critical point: $\rho \in \mathscr{V} \cap \mathcal{D}$; minimizes $\left|\rho_1 \cdots \rho_d\right|^{-1}$.
    - Smooth: $\nabla H(\mathbf{z}) \neq 0$ for all z in a neighbourhood of ρ.

- Output: $C_0, C_1, \ldots$ so that

$$f(r_1\, n,\ldots,r_d\, n) = \left(\rho_1^{r_1} \cdots \rho_d^{r_d}\right)^{-n}\left(\sum_{k=0}^{N-1} C_k n^{-(d-1)/2-k} + O(n^{-(d-1)/2-N})\right)$$

- Process

    - Express the coefficient as a multi-dimensional Cauchy integral.
    - Take a difference of Cauchy integrals so that ρ is the only singularity of the integrand inside the curve. This is possible by the strict minimality.
    - The coefficient is then expressed as a residue and an error term. The error has strictly smaller exponential growth.
    - The residue can be expressed as a Fourier-Laplace integral, using the variable transform $(\mathbf{z}) \mapsto (\rho_1 e^{it_1},\ldots,\rho_d e^{it_d})$.
    - Use approximation results for Fourier-Laplace integrals to estimate the coefficient.

---

handle a large number of word, and lattice walk problems. Equally important, it also captures the flow of the argument for more general cases. We handle these near the end of the chapter. The scope is broadened to powers and multiple factors in the next chapter. This will cover some non-smooth cases.

7.2 Warm-up: Stirling's Approximation

One of the first asymptotic estimates that an enumerator encounters is **Stirling's approximation of $n!$**:

$$n! \sim n^n e^{-n}\sqrt{2\pi n}.$$

Stirling's Approximation

$$n! \sim n^n e^{-n} \sqrt{2\pi n}.$$

As many counting formulas involve binomials and multinomials, such a simple approximation is very useful to understand the large-scale behaviour of combinatorial sequences. Indeed recall from Chapter 3 that diagonals of rational functions are always binomial sums, and so let us investigate it a little closer.

The approximation can be deduced by estimating the gamma function, $\Gamma(x)$, which is directly related to $n!$, and also has an integral form:

$$n! = \Gamma(n+1) = \int_0^\infty x^n e^{-x}\, dx.$$

We saw in Chapter 4 to use a saddle-point approximation to treat such an integral. Roughly, we examine the point on the contour of integration that the integrand is maximized, and rewrite the integral to pull out the dominant contribution, with a controlled error subterm.

More precisely, we first write $x^n e^{-x} = e^{n \log x - x}$, as it is more transparent that the integrand is maximized at $x = n$. Next, we apply the change of variables $t = n + u$:

$$n! = \int_{-n}^\infty e^{n \log(n+u) - (n+u)}\, du \tag{7.1}$$

$$= e^{n \log n - n} \int_{-n}^\infty e^{n \log\left(1 + \frac{u}{n}\right) - u}\, du \tag{7.2}$$

$$= e^{n \log n - n} \int_{-n}^\infty e^{-\frac{u^2}{2n} + O(u^3/n^2)}\, du \tag{7.3}$$

$$\sim e^{n \log n - n} \frac{\sqrt{2n\pi}}{2} \tag{7.4}$$

We justify Eq. (7.3) by using the Taylor expansion of $\log(1+z)$:

$$\log(1+z) = z + z^2/2 + O(z^3),$$

and Eq. (7.4) by the standard integral formula $\int_{-n}^\infty e^{-x^2/c}\, dx = \frac{\sqrt{c\pi}}{2}$.

The final step requires a bit of additional justification to ensure that the error term is correct. This can be argued precisely using the pace at which the integrand shrinks.

Let us highlight a few techniques that will work in general:

- Writing an integral as a Gaussian integral permits a quick transition to a closed form expression;

- Replacing functions by their Taylor series is useful to create estimates.

- Such estimates are optimal when computed near a local maximum.

7.3 Fourier-Laplace Integrals

The warm-up offers insight into a class of problems known as **Fourier-Laplace integrals**. These are integrals of the form

$$\int_{\mathscr{N}} A(\mathbf{t})e^{-\lambda\phi(\mathbf{t})}dt_1\ldots dt_d,$$

with the functions A and ϕ analytic over their domain of integration, and $\mathscr{N}$ is some neighbourhood in $\mathbb{R}^d$. The function A is called the **amplitude** and the function ϕ is called the **phase**. If ϕ is real, then it is a **Laplace**-type integral, and if it is purely complex, then it is said to be **Fourier**-type integral. As we work somewhere in between, we shall refer to Laplace-Fourier-type integrals.

Integrals of this form appear frequently in the study of probability density functions. Given a probability generating function, they are used to determine distributions. We say more in the Discussion.

In the one-dimensional case, we can apply some of the strategies from the warm up: To evaluate $\int A(z)e^{-\lambda\phi(z)}\,dz$, we find a critical point z_0 which minimizes $\phi(z)$ (to maximize the integrand), and then, assuming that ϕ is smooth at z_0, we approximate $\phi(z_0 + z)$ with the leading term of its Taylor expansion at z_0. Since $\phi'(z_0) = 0$, this series expansion looks like

$$\phi(z) = \phi(z_0) + C_k(z - z_0)^k + \ldots$$

for some non-zero C_k (very often $k = 2$). We do not go into the details, but we can prove the following theorem.

Proposition 7.1. *Let ϕ and A be real analytic functions. Suppose additionally that ϕ has a strict minimum at z_0, $\phi(0) = 0$, $\phi(z) \sim C(z-z_0)^2$ and $A(z_0) \neq 0$. Then,*

$$\int A(z) e^{-\lambda \phi(z)} \, dz \sim \sqrt{\frac{2\pi}{n\phi''(z_0)}} e^{-n\phi(z_0)}. \tag{7.5}$$

We illustrate how to apply this to combinatorial problems in the next section.

7.4   Easy Inventory Problems

Suppose A and B are nonlinear, nonconstant polynomials with positive integer coefficients. The simple bivariate function

$$F(x,y) = \frac{A(x)}{1 - yB(x)} = \sum_{(k,\ell) \in \mathbb{N}^2} f_{k\ell} x^k y^\ell$$

is combinatorial, as the $f_{k\ell}$ are all in $\mathbb{N}$. Generating functions for sequences with the cardinality of the sequence tracked as a parameter are of this form. This is common in walk and word problems. Let us consider how to approximate f_{nn} (or some other diagonal) for large n.

We note that the singular variety of F is

$$\mathcal{V} = \{(x,y) \in \mathbb{C}^2 \mid 1 - yB(x) = 0\} = \left\{ \left(x, \frac{1}{B(x)}\right) \mid x \in \mathbb{C} \right\}.$$

In this case, the critical point equation gives a solution to $B(x) = xB'(x)$ on this variety. As this is a combinatorial problem, there is a positive real solution $x = \rho$ to the equation $xB'(x) = 1$. We deduce that

$$\limsup_{n \to \infty} f_{nn}^{\frac{1}{n}} = \frac{B(\rho)}{\rho}.$$

However, we can determine a more precise estimate. Consider the following manipulation:

$$[x^n y^n]A(x)y^n B(x)^n = [x^n]A(x)B(x)^n \tag{7.6}$$

$$= \frac{1}{2\pi i}\int_{|x|=\epsilon}\frac{A(x)B(x)^n}{x^{n+1}}dx \tag{7.7}$$

$$= \frac{1}{2\pi i}\int_{|x|=\rho}\frac{A(x)B(x)^n}{x^{n+1}}dx \tag{7.8}$$

$$= \frac{1}{2\pi i}\int_{t=-\pi}^{\pi}\frac{A(\rho e^{it})B(\rho e^{it})^n}{\rho^{n+1}e^{it(n+1)}}i\rho e^{it}dt \tag{7.9}$$

$$= \frac{\rho^{-n}B(\rho)^n}{2\pi}\int_{t=-\pi}^{\pi}A(\rho e^{it})\frac{B(\rho e^{it})^n}{B(\rho)^n}e^{-it(n+1)}dt \tag{7.10}$$

$$= \frac{\rho^{-n}B(\rho)^n}{2\pi}\int_{t=-\pi}^{\pi}A(\rho e^{it})e^{-n\phi(t)}dt, \tag{7.11}$$

with $\phi(t) = \log\frac{B(\rho)}{B(\rho e^{it})} + it$. At this point we recognize a Laplace type integral, and we verify the hypotheses of Proposition 7.1 with minimum point $= 0$. We verify that $\phi(0) = 0$, and $\phi'(0) = 0$ by construction because because B has a critical point at ρ. The function $A(\rho e^{it})$ is clearly analytic since A is a polynomial. The criteria that $A(\rho) \neq 0$ must be verified in a case by case manner. We deduce from the proposition the first-order asymptotic formula:

$$[x^n y^n]F(x,y) \sim \left(\frac{B(\rho)}{\rho}\right)^n\frac{A(\rho)}{\sqrt{2\pi n\phi''(0)}}.$$

Example 7.1 (Balanced binary words with no runs). We can complete the asymptotic analysis of the number of balanced binary words over $\{0,1\}$ such that no word has a run of 1s of length 3 or longer. This number is given by

$$a(n,n) = [x^n y^n]\sum_{j,k}a(j,k)x^j y^k = \frac{1+x+x^2}{1-y(1+x+x^2)}.$$

There is a minimal critical point at $\rho = (1,1/3)$. Applying the above formula with $A(x) = B(x) = 1+x+x^2$,

$$a(n,n) \sim 3\frac{3^n}{\sqrt{2\pi n}}.$$

◂

7.5 Generalizing the Strategy to Higher Dimensions

We move to higher dimensions by first generalizing the Cauchy integral formula.

7.5.1 Multivariate Cauchy Integral Formula

Theorem 7.2 (Multivariate CIF). *Fix d and let $\mathbf{z} = (z_1, \ldots, z_d)$. Suppose that $F(\mathbf{z}) \in \mathbb{Q}(\mathbf{z})$ is analytic at the origin, with Taylor series expansion $F(\mathbf{z}) = \sum_{\mathbf{n} \in \mathbb{N}^d} f(\mathbf{n}) \mathbf{z}^{\mathbf{n}}$ Then for all $n \geq 0$,*

$$f(k_1, \ldots, k_d) = \frac{1}{(2\pi i)^d} \int_T \frac{F(\mathbf{z})}{z_1^{k_1} \ldots z_d^{k_d}} \cdot \frac{dz_1 \ldots, dz_d}{z_1 \ldots z_d}, \tag{7.12}$$

where T is the torus $T(\epsilon) = T(\epsilon_1, \epsilon_2, \ldots, \epsilon_d)$ has each ϵ_j sufficiently small, such that $F(\mathbf{z})$ is analytic in the interior of $D(\epsilon)$, and is analytic on the boundary.

The proof follows from the standard Cauchy integral formula by induction on the number of variables.

7.5.2 A Formula for Fourier-Laplace Integrals

The next step is to generalize Proposition 7.1. It is out of the scope of this discussion to prove such a theorem, but with the intuition that we have developed in the univariate case, the interested reader should be able to follow the details as provided in Pemantle and Wilson.

The general dimension version is the following, essentially Theorem 7.7.5 in [H90]. It uses the Hessian matrix $\mathcal{H}$ of ϕ which is defined as the $d \times d$ matrix with has j, k^{th} entry $\frac{\partial^2}{\partial t_j \partial t_k} \phi(\mathbf{t})$.

Proposition 7.3. *Suppose that the functions $A(\mathbf{t})$ and $\phi(\mathbf{t})$ in d variables are smooth in a neighbourhood $\mathcal{N} \subset \mathbb{R}^d$ of the origin and that*

- $\phi(\mathbf{0}) = 0$;

- *the real part of $\phi(\mathbf{t})$ is nonnegative on $\mathcal{N}$;*

- *ϕ has a critical point at $\mathbf{t} = \mathbf{0}$, i.e., that $(\nabla \phi)(\mathbf{0}) = \mathbf{0}$, and furthermore that the origin is the only critical point of ϕ in $\mathcal{N}$;*

- *The Hessian matrix of ϕ at $\mathbf{t} = \mathbf{0}$ is non-singular.*

Then for any integer $M > 0$ there exists effective constants $C_0, \ldots, C_M$ such that

$$\int_{\mathcal{N}} A(\mathbf{t}) e^{-n\phi(\mathbf{t})} d\mathbf{t} = \left(\frac{2\pi}{n}\right)^{d/2} \det(\mathcal{H})^{-1/2} \cdot \sum_{k=0}^{M} C_k n^{-k} + O\left(n^{-M-1}\right). \quad (7.13)$$

The constant C_0 is equal to $A(\mathbf{0})$. Moreover, if $A(\mathbf{t})$ vanishes to order L at the origin then (at least) the constants $C_0, \ldots, C_{\lfloor \frac{L}{2} \rfloor}$ are all zero.

The constants C_k are given by the formula:

$$C_k = (-1)^k \sum_{\ell \leq 2k} \frac{D^{\ell+k}(A \times \underline{\phi}^\ell)(0)}{2^{\ell+k} \ell!(\ell + k)!}, \quad (7.14)$$

with

$$\underline{\phi}(\mathbf{t}) := \phi(\mathbf{t}) - \sum t_j t_k \frac{\partial^2 \phi(\mathbf{t})}{\partial t_j \partial t_k}, \quad (7.15)$$

where D is the differential operator

$$D := \sum_{j,k} (\mathcal{H}^{-1})_{j,k} \frac{\partial^2}{\partial t_j \partial t_k}.$$

7.5.3 How Not to Transform This Integral

The plan will be to start with the Cauchy integral and end at a Fourier-Laplace integral. How do we do this? The most obvious change of variables $(z_1, \ldots, z_d) \mapsto \left(\epsilon_1 e^{it_1}, \ldots, \epsilon_d e^{it_d}\right)$, converts the Cauchy integral representation (7.12) into

$$f(\mathbf{n}) = \frac{(\epsilon_1 \ldots \epsilon_d)^{-n}}{(2\pi)^d} \int_{(-\pi,\pi)^d} F\left(\epsilon_1 e^{it_1}, \ldots, \epsilon_d e^{it_d}\right) e^{-n(t_1 + \cdots + t_d)} dt_1 \ldots dt_d.$$

In this integral, the ϕ is too simple! It is linear, $\phi = t_1 + \cdots + t_d$, and has no critical point. The strategy will be to incorporate the singular behaviour coming from H into ϕ.

To estimate coefficients univariate meromorphic functions we started with the Cauchy integral, moved the contour past the singularity, and then expressed the coefficient as a sum of residue computations and an error term.

To consider the contribution from isolated minimal critical points, we build a neighbourhood around them and convert the Cauchy integral into a Fourier-Laplace integral over that neighbourhood.

7.6   Example: Simple Walks

Recall we used a reflection principle argument (Eq. (3.22)) to determine the following expression for the generating function for simple walks in the quarter plane that end anywhere:

$$\sum_{n\in\mathbb{N}}\sum_{(k,\ell)\in\mathbb{N}^2}\mathrm{walk}((0,0)\xrightarrow{n}(k,\ell))z^n \tag{7.16}$$

$$= \Delta\left[xy\frac{(x-x^{-1})(y-y^{-1})}{(1-zxyS(x^{-1},y^{-1}))(1-x)(1-y)}\right] \tag{7.17}$$

$$= \Delta\left[\frac{(1+x)(1+y)}{1-xyzS(\frac{1}{x},\frac{1}{y})}\right]. \tag{7.18}$$

We are now in a position to determine an expression for the asymptotic growth of $w(n) := \sum_{(k,\ell)\in\mathbb{N}^2}\mathrm{walk}((0,0)\xrightarrow{n}(k,\ell))$ as n tends to infinity. Set

$$F(x,y,z) = G(x,y)/H(x,y,z) = \sum_n f(k_1,k_2,k_3)x^{k_1}y^{k_2}z^{k_3}$$

with $G = (1+x)(1+y)$ and $H = (1-xyzS(x^{-1},y^{-1}))$. Then,

$$w(n) = f(n,n,n).$$

We find C,μ,α so that $w(n)\sim C\mu^n n^\alpha$.

7.6.1   Exponential Growth

On the boundary of convergence $\partial\mathcal{D}$ of the Taylor series expansion of $F(x,y,z)$, the following relation is true: if $x\neq 1$ or $y\neq 1$ then $|z| = \left|xyS\left(x^{-1},y^{-1}\right)\right|^{-1}$. We have that the exponential growth μ of $f(n,n,n)$

$$\mu = \limsup_{n\to\infty} f(n,n,n)^{1/n}$$

$$= \inf_{(x,y,z)\in\partial\mathcal{D}}\left|xyz\right|^{-1} = \inf_{(x,y,z)\in\partial\mathcal{D}}\left|\frac{xy}{xyS\left(x^{-1},y^{-1}\right)}\right|^{-1}$$

$$= \inf_{(x,y,z)\in\partial\mathcal{D}}\left|S(x^{-1},y^{-1})\right|. \tag{7.19}$$

The step set of the simple walks is symmetric under reflection across either access. Consequently, the inventory polynomial $S(x,y)$ is symmetric in x and $1/x$ and also in y and $1/y$. Such a Laurent polynomial is necessarily minimized at $x = 1$ and $y = 1$, but possibly also at other roots of unity. We conclude that $S(x^{-1}, y^{-1})$ minimized when $|x| = |y| = 1$. The minimum value occurs at $(1,1)$ and also $(-1,-1)$:

$$\mu = \inf_{(x,y,z)\in\partial\mathcal{D}} \left|S(x^{-1},y^{-1})\right| = S(1,1) = 4.$$

There are two points for which $\left|S(x^{-1},y^{-1})\right| = 4$: $(1,1,1/4)$ and $(-1,-1,-1/4)$. These are both smooth points:

$$\nabla H(1,1,1/4) = (-1,-1,4) \quad \nabla H(1,1,4) = (1,1,1/4)$$

We have argued directly, but note also that these are the solutions to the critical point equations determined via a Gröbner basis computation:

$$\text{basis}\{H, xH_x - zH_z, yH_y - zH_z\} = [16z^2 - 1, y - 4z, x - 4z].$$

Note in Eq. (7.17) the $(1 - x)$ and $(1 - y)$ terms in the denominator did cancel with terms in the numerator. Were this not true, then the singular variety is no longer smooth at the critical point $(1,1,1/4)$ since it is annihilated by all three factors.

The variety will not be smooth at a critical point whenever the critical point is annihilated by more than one factor. When this happens, the point is said to be a multiple point. We consider multiple points in the next chapter.

Interestingly, we note that up to Equation (7.19) the argument is valid for any inventory polynomial $S(x,y)$ of a lattice walk. Furthermore, a similar argument is true for simple walks in higher dimensions. Thus, we have relatively direct access to the exponential growth of lattice walks in higher dimension provided we can express the generating function as a diagonal.

7.6.2 Estimating Cauchy Integrals

Now we refine the estimate of the coefficient growth. We have

$$f(n,n,n) = [x^n y^n](1 + x)(1 + y)(xyz)^n S(x^{-1}, y^{-1})^n = [x^0 y^0]G(x,y)S\left(x^{-1}, y^{-1}\right)^n$$

$$= \iint_{T(\epsilon_1,\epsilon_2)} G(x,y)S(x,y)^n \frac{dx\,dy}{xy} \tag{7.20}$$

for a small torus around the origin. Note, that we have applied to simplification $S(1/x, 1/y) = S(x,y)$ here. We can expand the contour out as far as needed, as $G(x,y)S(x,y)^n$ is analytic. Expand the torus of $(1,1)$:

$$f(n,n,n) = \frac{1}{(2\pi i)^2} \iint_{T(1,1)} G(x,y)S(x,y)^n \frac{dx\,dy}{xy} \qquad (7.21)$$

We use the variable substitution

$$x \mapsto \rho_1 e^{it_1} \quad y \mapsto \rho_2 e^{it_2} \quad dx \mapsto ie^{it_1}dt_1 \quad dy \mapsto ie^{it_2}dt_2$$

and rewrite Eq. (7.21):

$$f(n,n,n) \qquad (7.22)$$

$$= \frac{(i)^2}{(2\pi i)^2} \iint G(\rho_1 e^{it_1}, \rho_2 e^{it_2})S\left(\rho_1 e^{it_1}, \rho_2 e^{it_2}\right)^n dt_1 dt_2 \qquad (7.23)$$

$$= \frac{1}{(2\pi)^2} \iint G(\rho_1 e^{it_1}, \rho_2 e^{it_2})S\left(\rho_1 e^{it_1}, \rho_2 e^{it_2}\right)^n \frac{S(\rho_1, \rho_2)^n}{S(\rho_1, \rho_2)^n} dt_1 dt_2 \qquad (7.24)$$

$$= \frac{S(\rho_1, \rho_2)^n}{(2\pi)^2} \iint G(\rho_1 e^{it_1}, \rho_2 e^{it_2})e^{-n\phi(t_1, t_2)} dt_1 dt_2 \qquad (7.25)$$

where

$$\phi(t_1, t_2) = \log\left(\frac{S(\rho_1, \rho_2)}{S(\rho_1 e^{it_1}, \rho_2 e^{it_2})}\right).$$

Now, let us verify the hypotheses of Proposition 7.3. Here $\mathcal{N} = (-\pi, \pi)^2$, and ϕ and G are smooth in this domain. It is easy to verify that $\phi(0,0) = 0$, and that $\nabla\phi(0) = \partial_1\phi(0,0) + \partial_2\phi(0,0) = 0$. The real part of $\phi(t_1, t_2)$ is $\log\left(\frac{S(\rho_1, \rho_2)}{S(\rho_1 e^{it_1}, \rho_2 e^{it_2})}\right)$, and this is nonnegative as $S(x,y)$ is maximized locally in this domain at (ρ_1, ρ_2) maximizes value.

However, there is a critical flaw: The origin is not the **only** critical point in the region! As $S(x,y)$ is also minimized at $(-1,-1)$, the point $(-\pi, -\pi)$ is a second critical point for ϕ. We fix this problem by shifting our parametrization of the domain and then splitting it into two parts: $\mathcal{N} = \mathcal{N}_1 \cup \mathcal{N}_2$ with

$$\mathcal{N}_1 = (-\pi/2, 3\pi/2) \times (-\pi/2, \pi/2), \quad \mathcal{N}_2 = (-\pi/2, 3\pi/2) \times (\pi/2, 3\pi/2).$$

We repeat the above process to get

$$(2\pi i)^2 f(n,n,n) = \int_{\mathcal{N}_1} G(x,y) S(x,y)^n \frac{dx\,dy}{xy} + \int_{\mathcal{N}_1} G(x,y) S(x,y)^n \frac{dx\,dy}{xy}$$

$$= S(1,1)^n \int_{\mathcal{N}_1} G(e^{it_1}, e^{it_2}) e^{-n\phi_1(t_1,t_2)} dt_1 dt_2$$

$$+ S(-1,-1)^n \int_{\mathcal{N}_2} G(e^{it_1}, e^{it_2}) e^{-n\phi_2(t_1,t_2)} dt_1 dt_2.$$

where

$$\phi_1(t_1,t_2) := \log\left(\frac{4}{S(e^{it_1}, e^{it_2})}\right), \text{ and}$$

$$\phi_2(t_1,t_2) := \log\left(\frac{-4}{S(-e^{it_1}, -e^{it_2})}\right),$$

The contribution from $(1,1,1/4)$ is $\frac{4^{n+1} n^{-1}}{\pi}$ whereas the dominant term from $(-1,-1,-1/4)$ contributes a lower order polynomial in the subexponential term. We conclude that

$$f(n,n,n) \sim \frac{4^{n+1} n^{-1}}{\pi}.$$

7.7 A More General Strategy

We can solve a more general class of problems using this solution as a template: The coefficient is expressed as a Cauchy integral. The solution is a sum over contributions indexed by minimal critical points. The contribution from each critical point is computed in parallel: First, the variety is parametrized locally around that critical point. The function is rewritten as a function in lower dimension, and the innermost integral is computed as a residue. The result is transformed into an integral of Fourier-Laplace type where the neighbourhood $\mathcal{N}$ in the hypothesis is determined by the parametrization, but by construction will only contain one critical point. We have seen the parametrization $z_{d+1} = \frac{1}{B(z_1,\ldots,z_d)}$ from the previous example when $H = 1 - z_{d+1} B(z_1,\ldots,z_d)$. In the last case, the substitution immediately reduced the dimension of the problem, but the general case is more subtle. Let us consider it now.

Summary

Input: $A, B \in \mathbb{N}[\mathbf{z}]; \mathbf{r} = (r_1, \ldots, r_d)$.
Output: C, μ, α so that

$$[z_1^{nr_1} \cdots z_d^{nr_d}] A(\mathbf{z}) B(\mathbf{z})^n \sim C \mu^n n^{\alpha} \text{ as } n \to \infty$$

Step 1: Determine potential contributing critical points

- Find the critical points; the solutions to the critical point equations with $H = (1 - z_{d+1} B(\mathbf{z}))$:

$$H = 0, \quad r_1^{-1} z_1 H_1 = r_k^{-1} z_k H_k, \quad k = 2, \ldots, d+1.$$

- Find minimal critical points: Exclude those critical points ρ such that $T(\rho) \not\subset \partial \mathcal{D}$.

- Exponential growth: Set

$$\mu = \min_{\rho} \left| \rho_1^{r_1} \cdots \rho_d^{r_d} \rho_{d+1} \right|^{-1} = \min_{\rho} \frac{\rho_1^{r_1} \cdots \rho_d^{r_d}}{B(\rho_1, \ldots, \rho_d)}.$$

- Let $\mathcal{C}$ be the set of minimal critical points that attain this minimum.

Step 2: For each of the critical points ρ in $\mathcal{C}$, determine the contribution $\Psi_{\rho}(n)$ to the asymptotics.

- Verify smoothness condition, $\nabla H(\rho) \neq \mathbf{0}$. If not smooth, then STOP. This might be a multiple point – see next chapter.

- Let $\phi(\mathbf{t}) = \log(B(\rho)/B(\rho e^{i\mathbf{t}})) + i(r_1 t_1 + \cdots + r_{d-1} t_{d-1})$

- Compute $\det \mathcal{H} = \det[\partial_i \partial_j \phi''(0)]$. If this is zero, STOP. (You might be working in the wrong dimension.)

- Incrementally compute $C_0, C_1, \ldots$, according to Eqn. (7.14). Let N be the smallest index for which $C_N \neq 0$. Note, if $A(\rho) \neq 0$, $N = 0$, and

$$C_0 = \sqrt{\pi \det(\mathcal{H})}^{-1} A(\rho).$$

- $\Psi_{\rho}(n) = C_N \mu^{-n} n^{d/2 - N}$

Step 3: Sum over contributions.

$$f(n\mathbf{r}) \sim \sum_{\rho \in \mathcal{C}} \Psi_{\rho}(n)$$

The approximation is accurate for the dominant term.

Recall that a point in the variety $\mathscr{V} = \{\mathbf{z} \mid H(\mathbf{z}) = 0\}$ is minimal if and only if $D(\mathbf{z}) \cap \mathscr{V} \subset \partial D(\mathbf{z})$. That is, the only points on the polydisk of $\mathbf{z}$ that are also in the variety must occur on the boundary of the polydisk. If the set of such points is finite, the point is finitely minimal. If $\mathbf{z}$ is the only one, then it is a strictly minimal point. The fact that our singularities are finitely minimal is crucial to the argument below.

For now let us first suppose that $F(\mathbf{z})$ has a strictly minimal point ρ. (If they are finite in number we will build a sum as in the previous example.) If the variety $\mathscr{V}$ is smooth near this point, and if $\partial_{d+1} H(\rho) \neq 0$, then the implicit function theorem says there exists a parametrization that we can use. Specifically, there is a bounded neighbourhood W of a point $\tilde{\rho}$, and a holomorphic function ψ so that $(\mathbf{w}, \psi(\mathbf{w})) \in \mathscr{V}$, and $\partial_d H(w, \psi(w)) \neq 0$ for all $\mathbf{w} \in W$ and $\psi(\tilde{\rho}) = \rho_d$. We consider critical points with no component equal to zero, so notably $\rho_d \neq 0$, hence we can assume ψ is non-zero on W. In the case $H = 1 - z_{d+1} P(\mathbf{z})$ this is precisely what we had before: $\psi(\mathbf{z}) = \frac{1}{P(\mathbf{z})}$.

Now, in the case before we simplified the problem so that the integrand to the Cauchy integral had no singularities away from the origin. This allowed us to choose the contour we needed. In general we need to be more subtle and use an argument more reminiscent of the proof of Theorem 4.11 for meromorphic asymptotics. Recall in that proof a contour containing singularities in its interior is built and is then rewritten as a sum of contours each containing exactly one pole of the function, and also a contour containing the origin. Rearranging this equation gave an expression for a coefficient as a sum of residue integrals around each pole, plus an error term of exponentially smaller growth. It is worth revisiting before continuing on, if this summary does not jog your memory.

We can do something similar here. Let us lean into the analogy, and then then details. In the smooth case, we can do something similar. Let ρ be a strictly minimal critical point of $F(\mathbf{z}) = G(\mathbf{z})/H(\mathbf{z})$, with $\mathscr{V}$ smooth at ρ. Let T^* be the $d-1$ dimensional torus $T(\rho_1, \ldots, \rho_{d-1})$.

The analog of the circle around the origin is a torus $T(\epsilon)$, such that F is analytic on its interior:

$$f(n, \ldots, n) = \left(\frac{1}{2\pi i}\right)^d \int_{T(\epsilon)} \frac{F(\mathbf{z})}{(z_1 \ldots z_d)^n} \frac{dz_1 \ldots dz_d}{z_1 \cdots z_d} \tag{7.26}$$

$$= \left(\frac{1}{2\pi i}\right)^d \int_{T(\rho_1, \ldots, \rho_{d-1})} \int_{|z_d| = \rho_d - \delta} \frac{F(\mathbf{z})}{(z_1 \ldots z_d)^n} \frac{dz_1 \ldots dz_d}{z_1 \cdots z_d}. \tag{7.27}$$

In this new domain F remains analytic in the interior.

The analog of the circle around ρ is two torii defined using some $\delta > 0$: $T_1 = T(\rho^-, \rho_d + \delta)$ and $T_1 = T(\rho^-, \rho_d - \delta)$. Thus, ρ is in the interior of the two torii:

$$f(n, \ldots, n) = \underbrace{\int_{T_1} - \int_{T_2}}_{\text{simplify}} + Err.$$

We can reduce the dimension by evaluating the interior-most integral using a simple residue computation once we parametrize the function. By the implicit function theorem, there is a neighbourhood $\mathcal{N}$ of $\rho^- := (\rho_1, \ldots, \rho_{d-1})$, and an analytic function $\psi : \mathcal{N} \to \mathbb{C}$, such that $H(\rho^-, \psi(\rho^-)) = 0$ for all $z \in \mathcal{N}$. Remarkably, it can be shown additionally,

$$\left|\rho_d\right| \leq \left|\psi(z_1, \ldots, z_{d-1})\right| < \left|\rho_d\right| + \delta$$

for $\mathbf{z}^- \in \mathcal{N}$, with the lower equality occurring only when $\mathbf{z}^- = \rho^-$. Finally, $H(\mathbf{z}^-, w) \neq 0$ if $w \neq \psi(\rho^-)$ and $|w| < \rho_d$:

$$f(n, \ldots, n) = \left(\frac{1}{2\pi i}\right)^d \int_{\mathcal{N}} \left(\int_{|z_d| = \rho_d + \delta} F(\mathbf{z}^-, z_d) \frac{dz_d}{z_d^{n+1}}\right) \frac{dz_1 \ldots dz_{d-1}}{(z_1 \cdots z_{d-1})^{n+1}}.$$
$$(7.28)$$

For each $\mathbf{z}^- \in \mathcal{N}$, $F(\mathbf{z}^-, z_d)$ has a **unique singularity** between $|z_d| = \rho_d - \delta$ and $|z_d| = \rho_d + \delta$, a *simple pole* at $z_d = \psi(\mathbf{z}^-)$. For any fixed $\mathbf{z}^-$, let $\mathcal{N}'$ be the closed contour composed by joining the clockwise circle of radius $\left|\rho_d\right| - \delta$ and the counter-clockwise circle of radius $\left|\rho_d\right| + \delta$. When we examine the interior integral above, we can replace it with a simple residue, using the formula for rational functions $Res_{z=z_0} P(z)/Q(z) = 2\pi i P(z_0)/Q'(z_0)$:

$$\frac{1}{2\pi i} \int_{\mathcal{N}'} F(\mathbf{z}^-, z_d) \frac{dz_d}{z_d^{n+1}} = \frac{G(\mathbf{z}^-, \psi(\mathbf{z}^-))}{\psi(\mathbf{z}^-)^{n+1} \frac{\partial H}{\partial z_d}(\mathbf{z}^-, \psi(\mathbf{z}^-))}.$$

Once we do that we are left the following expression:

$$f(n, \ldots, n) \sim \int_{T_1} - \int_{T_2} = \frac{1}{(2\pi i)^{d-1}} \int_{\mathcal{N}} \frac{-G(\mathbf{z}^-, \psi(\mathbf{z}^-))}{\psi(\mathbf{z}^-)^{n+1} H_{z_d}(\mathbf{z}^-, \psi(\mathbf{z}^-))}. \quad (7.29)$$

Inside $\mathcal{N}$ there is exactly one critical point by construction. To translate this expression into a Fourier-Laplace–type integral, use the following ϕ and proceed as in the last section:

$$\phi(\boldsymbol{\theta}) = \log(\psi(\rho_1 e^{it_1}, \ldots, \rho_{d-1} e^{it_{d-1}})) - \log(\rho) + \frac{i}{r_d}(r_1 t_1 + \cdots + r_{d-1} t_{d-1}).$$

The A is the image of

$$\frac{-G(\mathbf{z}^-, \psi(\mathbf{z}^-))}{H_{z_d}(\mathbf{z}^-, \psi(\mathbf{z}^-))}$$

under $z_i \mapsto \rho_i e^{it_i}$.

Theorem 7.4. *Let $F(z)$ be a rational function with a square-free denominator, which is analytic at the origin and has a smooth singular variety $\mathcal{V}$. Assume that F admits a quadratically nondegenerate, strictly minimal critical point ρ, and that $\frac{\partial H}{\partial z_d}(\rho) \neq 0$. Then, for any nonnegative integer M,*

$$f(n, \ldots, n) = \frac{(\rho_1 \cdots \rho_d)^{-n}}{n^{(d-1)/2} (2\pi)^{(d-1)/2} \det \mathcal{H}^{1/2}} \left(\sum_{j=0}^{M} C_j n^{-j} + O(n^{-M-1}) \right),$$

where the C_i are computed in Equation (7.14). The leading constant C_0 in this series has the value

$$C_0 = \frac{-G(\rho)}{\rho_n \frac{\partial H}{\partial z_n}(\rho)}.$$

This C_0 is defined and non-zero when $G(\rho) \neq 0$.

We have not addressed here to how to find such a ψ, which is clearly needed to define ϕ, and hence apply the above formulas. Sometimes it is possible to directly solve for y. In two dimensions, it can be straightforward, but you need to ensure that you are choosing the right solution in case of multiple solutions. The article [RW11] provides strategies and examples and does permit some explicit expressions.

Theorem 7.5. *Let $F(\mathbf{z}) = G(\mathbf{z})/H(\mathbf{z})$ be meromorphic and suppose that as $\hat{\mathbf{r}}$ varies in a neighbourhood N of $\mathbf{r}$, there is a smoothly varying, strictly minimal smooth critical point ρ in the direction $\mathbf{r}$. Suppose also that $G(\rho) \neq 0$. Define $\mathbf{z}(\mathbf{r})$ as the critical point in the direction $\mathbf{r}$, and define*

$$Q := -y^2 H_y^2 x H_x - y H_y x^2 H_x - x^2 y^2 \left(H_y^2 H_{xx} + H_x^2 H_{yy} - 2 H_x H_y H_{xy} \right).$$

If this function is non-zero in a neighbourhood of $\mathbf{z}(n\mathbf{r})$ then

$$f(n\mathbf{r}) \sim \frac{G(\rho)\left(\rho_1^{-r}\rho_2^{-s}\right)^n}{\sqrt{2\pi}} \sqrt{\frac{-\rho_2 H_y(\rho)}{ns Q(\rho)}},$$

as n tends to infinity.

Now, one must be careful when taking the square root to use the appropriate branch.

Example 7.2 (Delannoy). We apply Theorem 7.5 to determine the asymptotic growth of Delannoy numbers:

$$f(nr, ns) \sim \left(\frac{r}{\sqrt{r^2 + s^2} - s} \right)^n \left(\frac{s}{\sqrt{r^2 + s^2} - r} \right)^n \sqrt{\frac{nrs}{2\pi\sqrt{r^2 + s^2}}} \frac{1}{r + s - \sqrt{r^2 + s^2}}.$$

◁

Writing comparable theorems in higher dimensions becomes quite messy. Instead, one can use alternative formulas which incorporate the **Gaussian curvature** of the singular variety. This is out of scope for this text. See [PW13, Section 9.5].

7.8   Discussion

Bertozzi and McKenna [BM93] introduced the use of multidimensional residues to determine asymptotic estimates of generating function coefficients. An important consequence of these computations is the **central limit theorem**. This is a major result which proves that data influenced only by many small and unrelated random effects are approximately normally distributed. One can prove the theorem in a manner analogously to the integral approximations we have done above:

$$P(x) = \sum p_n x^n \implies p_n \sim \frac{1}{2\pi} \int e^{-x} P(e^x) dx.$$

Drmota [Drm94] provides an excellent reference for some of the finer manipulations in the combinatorial case of the type in Section 7.5.

Random walks in cones provides numerous examples of this type, and Biane's formulas for walks in Weyl chambers [Bia93, Bia92] provides estimation of a similar nature, albeit with formulas phrased for the expert. The explicit generating function formulations for Weyl chamber walks written plainly as constant terms of rational functions are provided by Zeilberger [Zei83] and Gessel and Zeilberger [GZ92]. The asymptotic analysis of these models is initiated by Grabiner [Gra06, Gra02], and Krattenthaler [Kra07] derives a direct saddle-point approximation. The mechanics of the argument have some common elements to some of the estimates here.

Theorems of Hörmander are applied by Tate and Zelditch [TZ04] to determine asymptotic expressions for walks in Weyl chambers and elaborated to other convex cones by Tate [Tat11] and Feierl [Fei10]. These are presented in a functional analysis context.

Quantum random walks in this context are the topic of Example 9.5.5 in [PW13]. It is presented in context and discusses the underlying geometry.

Raichev developed a Sage package which serves as a companion to the article [Rai12a] which determines the asymptotic contribution of a non-degenerate critical point on the border of the domain of convergence at which $\mathcal{H}$ is smooth or the transverse intersection of smooth algebraic varieties. This is useful if you already know which critical points contribute.

The main source for the smooth case is the text of Chapter 9 or Pemantle and Wilson [PW13], and the references therein specifically Corollary 9.2.3 and Theorem 9.2.7. It is their fine proofs which make the smooth case cleanly presented and easy to use.

The presentation of Melczer [Mel17], which contains simple walks analyzed by Melczer and Mishna [MM16], is done in sufficient detail to be complete, and instructive.

7.9 Problems

Exercise 7.1. Formalize the proof of Stirling's approximation. You will need to explain precisely why most of the terms in the series expansion of the logarithm do not contribute to the dominant asymptotics. Compute the second term to determine a more precise formula. ❑

Exercise 7.2. Determine the asymptotic growth of $[x^n y^n]\frac{1}{1-yB(x)}$ when $B(x) = ax^2 + bx + c$. Verify explicitly that $\phi'(0) = 0$, provided that B is at least quadratic. What goes wrong when B is linear? Under which conditions is $A(\rho) = 0$? What does this mean, and how can should one proceed? ❑

Exercise 7.3. Redo the binary words example with arbitrary direction vector **r**. ❑

Exercise 7.4. Consider the formal language

$$\mathscr{L} = \{\text{binary expansions of } n \mid n \equiv 0 \mod 3\}.$$

The size of a word is the length of the string. This set looks like $\mathscr{L} = \{\epsilon, \overset{0}{0}, 00, 000, \ldots, \overset{3}{11}, 011, 0011, \ldots, \overset{6}{110}, 0110, 00110, \ldots,$
$\overset{9}{1001}, 01001, \overset{12}{1100}, 01100, \ldots, \overset{15}{1111}, 01111, \ldots\}$.

1. Show that $\mathscr{L}$ satisfies the S-regular specification: $\mathscr{L} = (0 + (1(01^*0)^*1))^*$.

2. Show that the parameter $\chi(w) = (|w|_0, |w|_1) = (\#0s \text{ in } w, \#1s \text{ in } w)$ is an inherited parameter.

3. Write the generating function for the derived class of balanced words as a diagonal

$$\mathcal{L}_= = \{w \in \mathcal{L} \mid \chi(w) = (n, n), n \geq 0\}$$
$$= \{w \in \mathcal{L} \mid \#0s = \#1s\}$$
$$= \{\mathbf{1001}, 0011, 0110, 1100, \mathbf{010101}, \mathbf{101010}, \mathbf{11100001}, \mathbf{10011001},$$
$$\mathbf{10000111}, 00101101, \mathbf{01011010}, \mathbf{00111001}, \mathbf{00100111}, \ldots\}$$

4. Determine a formula for the asymptotic number of balanced words in this language of length $2n$.

❑

Exercise 7.5. Prove the formula

$$\lim f(\mathbf{n})^{1/n} \leq |\rho_1 \cdots \rho_d|^{-1}$$

bounding the exponential growth factor by bounding the Cauchy integral. Hint: $\left| \int_\gamma f \right| \leq \max z \in \gamma |f(z)| \operatorname{length} \gamma.$ ☐

Exercise 7.6. Let $\mathcal{E}$ be the set of walks with steps $\pm e_1, \pm e_2$ in the plane that start and end at $(0,0)$.

1. Show that the generating function is equal to $\Delta \frac{A(x,y)}{1-zB(x,y)}$ with $A(x,y) = 1$ $B(x,y) = xy\left(x + \frac{1}{x} + y + \frac{1}{x}\right)$.

2. Prove that there are two critical points ρ: $(1,1,1/4), (-1,-1,-1/4)$.

3. Prove that the number of excursions of length n, denoted $e(n)$ satisfies

$$e(n) \sim \frac{4^n}{\pi n} + \frac{(-4)^n}{\pi n}.$$

☐

8

Multiple Points

CONTENTS

8.1   Algebraic Geometry Basics 178
8.2   Critical Points 180
8.3   Examples 182
       8.3.1   Tandem Walks 182
       8.3.2   Weighted Simple Walks 184
8.4   A Direct Formula for Powers 185
8.5   The Contribution of a Transverse Multiple Point 186
8.6   Discussion 188
8.7   Problems 189

Estimating the coefficient behaviour of the diagonals of the rational function $G(\mathbf{z})/H(\mathbf{z})$ can be straightforward when H is irreducible, the critical points are finitely minimal, and the singular variety $\mathcal{V} = \{\mathbf{z} \mid H(\mathbf{z}) = 0\}$ is smooth at the contributing critical points as we saw in the previous chapter.

What can change if H factors? Some additional care is required in the search for critical points and their subsequent processing. Products of distinct factors, and powers of a single factor give rise to different complications. Powers of a factor do not change the singular variety, but they do change the contribution – we have seen this in the transfer theorem in the univariate case. We are lucky when H has multiple factors if near the contributing critical points the singular variety decomposes as a union of smooth varieties. A critical point in an intersection of multiple varieties is said to be a **multiple point**.

Our estimates are still built as sums indexed by contributing critical points. In this chapter, we refine the process to find the critical points, and see the formulas for $\Phi_\rho(n)$, the contributions of a multiple point to the dominant asymptotics of the coefficient sequence. Developing the formulas is out of scope here.

We must start with some basic geometry terminology so we can describe how to determine the critical points. The definitions are illustrated with lattice walk enumeration problems. This is followed by some explicit formulas for the contributions, which are again essentially estimates of residue integrals. The reader is cautioned that for ease of presentation we make numerous simplifying assumptions. For example, we are unapologetic about the choice to consider only problems with finitely minimal critical points.

8.1 Algebraic Geometry Basics

Consider the rational function $F(\mathbf{z}) = \frac{G(\mathbf{z})}{H(\mathbf{z})}$ and its singular variety $\mathscr{V} = \mathscr{V}_H = \{\mathbf{z} \in \mathbb{C}^d : H(\mathbf{z}) = 0\}$. Suppose H factors nontrivially into square-free factors

$$H = H_1^{m_1} H_2^{m_2} \cdots H_k^{m_k}, \quad m_i \in \mathbb{Z}_{>0}.$$

In this case, the variety $\mathscr{V}_H$ decomposes into a union of subvarieties:

$$\mathscr{V}_H = \mathscr{V}_{H_1} \cup \cdots \cup \mathscr{V}_{H_k}.$$

A polynomial and a non-zero positive integer power of that polynomial have the same roots. In terms of finding critical points, and understanding the geometry of $\mathscr{V}$, much of the general discussion remains the same once we remove the repeat factors from H. For example, if the critical point is only a root of a single factor, say H_1, and the exponent of this factor is one, e.g., $m_1 = 1$, then provided the variety $\mathscr{V}_{H_1}$ is smooth at that critical point, then we view F as a meromorphic function, divided by a polynomial:

$$\frac{G\left(H_2^{m_2} \cdots H_k^{m_k}\right)^{-1}}{H_1} = \frac{\hat{G}}{H_1}$$

when we process that point, which permits us to reduce to the smooth case of the previous chapter.

If a manifold (or other reasonable space) can be decomposed into connected submanifolds with strictly diminishing dimension, then we say it can be **stratified**, and each component is a **stratum**. Sets defined by the vanishing of finitely many polynomials can be so stratified. The strata satisfy the **axiom of the frontier**: The closure of each stratum should be a union of lower dimensional strata. In our applications, we

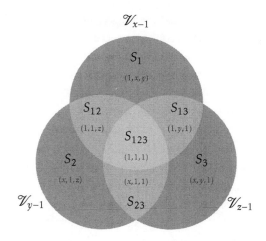

FIGURE 8.1
Venn diagram schematic of the stratification of $\mathcal{V}_{(1-z)(1-x)(1-y)}$.

will stratify $\mathcal{V} \cap \partial D$, the intersection of the variety and the domain of convergence of the series expansion we study.

A point $z \in \mathcal{V}$ is a **multiple point** if, for every sufficiently small neighbourhood U of z in $\mathcal{V}$, the set U is the union of finitely many smooth varieties. By definition, any point is an element of a unique stratum of $\mathcal{V}$. Let us consider the stratum containing the multiple point ρ in the singular variety of F. If the co-dimension of the stratum containing ρ is k, then we say that **the order of the intersection at ρ is k**.

Example 8.1. Let $H(x,y,z) = (1-x)(1-y)(1-z)$, and let the manifold in consideration be the real variety $\mathcal{V}_H \subset \mathbb{R}^3$. It is decomposed into seven strata represented in Figure 8.1. For example $S_{12} = \mathcal{V}_{x-1} \cap \mathcal{V}_{y-1} \setminus S_{123}$. The closure of this stratum is the line $z = 1$, which is of dimension 1, and hence of co-dimension 2. Any point in S_{12} is of order 2. ◄

Multiple points introduce new possibilities for the geometry at the singular point. In this chapter, we limit ourselves to the case of transversal multiple points (detailed in the next section). In the next chapter, we consider a combinatorially motivated situation that can be systematically reduced to this case.

Intuitively, two curves have a **transversal intersection** if the intersection is robust to small perturbations of the curves. For example, the curves $y = x^2$ and $y = 0$ do *not* have a transversal intersection at $(0,0)$ because at that point they meet exactly once, but a small perturbation of either curve, say, $y = x^2 + \epsilon$ results in either two or no intersection points. More formally, a multiple point $\mathbf{z} \in \mathcal{V}$ is a **transversal multiple point** if in addition to the properties of a multiple point, the normal vectors to the $\mathcal{V}_j$ at $\mathbf{z}$ are linearly independent. In the example, we see the normals to the curves $y = x^2$ and $y = 0$ are in the same direction at $(0,0)$; they are both along the x axis.

To detect a transversal intersection we can test the normal characterization. This implies that in dimension d, the intersection of more than d varieties cannot be transversal. When X and Y meet transversely, then $X \cap Y$ is smooth, and of dimension $\dim X + \dim Y - d$. Two curves that do not intersect transversely might be well behaved but be of too small dimension or it might be a pathological – such as at a cusp.

8.2   Critical Points

Recall that a critical point of F minimizes $\left| z_1^{r_1} \dots z_d^{r_d} \right|^{-1}$ for $\mathbf{z}$ in the closure of the domain of convergence. It is a minimizer for the height function: This is the smallest real constant c such that the hyperplane $r_1 z_1 + \dots + r_d z_d = c$ is tangent to the curve $\{-\log|z_1| - \dots - \log|z_d| \mid \mathbf{z} \in \partial \mathcal{D}\}$. The point of intersection under the relog map, in $\mathbb{R}^d$, may represent several critical points in $\mathbb{C}^d$. These critical points lead to an exponential growth factor of $\left| z_1^{r_1} \dots z_d^{r_d} \right|^{-1}$. To efficiently find critical points, we examine the subset of $\mathcal{V}$ that intersects the boundary of convergence, $\partial \mathcal{D}$. This may involve searching for critical points in each strata.

Let us first develop the intuition by considering what happens when $d = 2$ and the critical points are isolated. Suppose $H = H_1 H_2$. Figure 8.2 is a simplified representation of a possible boundary of convergence, where we see a segment contributed by the image of $\mathcal{V}_{H_1}$, a second by the image of $\mathcal{V}_{H_2}$, and their point of intersection. The minimizing process finds the smallest c such that the line $r\hat{x} + s\hat{y} = c$, where $\hat{x} = -\log|x|$ and $\hat{y} = -\log|y|$. touches this interface. It may interact with both boundaries as it appears to in the figure. It is necessary to

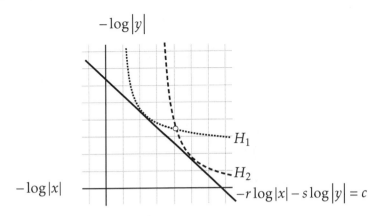

FIGURE 8.2
The line $r\hat{x} + s\hat{y} = c$ touches the boundary of -relog $\mathscr{V}$ =-relog $\mathscr{V}_{H_1 H_2}$ in two distinct places. Each intersection corresponds to a distinct stratum. The small circle is the image under -relog of a transversal intersection point.

compare the value $\left| x^r y^s \right|^{-1}$ for every intersection to find the minimizing points. This process generalizes naturally to higher dimensions.

To summarize: given the intersection $\mathscr{V} \cap \partial \mathcal{D}$, we find a stratification, and look for candidate critical points from each stratum. We select those that minimize $\left| z_1^{r_1} \ldots z_d^{r_d} \right|^{-1}$ using the critical point equations as before using only the relevant factors from H. To determine the contribution of the critical point to the coefficient asymptotics, we examine the geometry at that point. If it is on a single variety, then it is probably a smooth point. If it is in the intersection of more than one variety, it is a multiple point. There is a straightforward algebraic test to determine whether or not a multiple point is a transverse multiple point.

Proposition 8.1 (Critical point criteria). *Suppose that S is a stratum of $\mathscr{V}$ defined as $S = \mathscr{V}_{H_1 \ldots H_k} \setminus \mathcal{A}$, where $\mathcal{A}$ is an algebraic set of lower dimension than $\mathscr{V}_{H_1 \ldots H_k}$; each $\mathbb{V}(H_j)$ is a complex manifold; and the tangent planes of the $\mathscr{V}_{H_j}$ are linearly independent where they intersect. Then, $\rho \in S$ is a critical point if and only if the vector $\mathbf{r}$ can be written as a linear combination of the vectors*

$$\left(z_1 \frac{\partial H_i}{\partial z_1}, z_2 \frac{\partial H_i}{\partial z_2}, z_d \frac{\partial H_i}{\partial z_d} \right), \qquad i = 1, \ldots, r.$$

TABLE 8.1
Stratification of $\mathscr{V} \cap \partial \mathcal{D}$

Stratum	Description	Critical point ρ	$\lvert \rho_1 \rho_2 \rho_3 \rvert^{-1}$
S_1	$\{(x,y,\frac{1}{y+xy^2+x^2}) \mid x \neq 1, y \neq 1\}$	$(\omega, \omega^2, \omega^2/3)$	$1/3$
		$(\omega^2, \omega, \omega/3)$	$1/3$
S_{12}	$\{(x,1,\frac{1}{1+x+x^2}) \mid x \neq 1\}$		
S_{23}	$\{(1,y,\frac{1}{1+y+y^2}) \mid y \neq 1\}$		
S_{123}	$\{(1,1,1/3)\}$	$(1,1,1/3)$	$1/3$

Proposition 8.2. *The point $\rho \in \mathscr{V}$ is a multiple point if and only if there is a factorization*

$$H = \prod_{j=1}^{N} H_j^{m_j},$$

with $\nabla H_j(\rho) \neq 0$ and $H_j(\rho) = 0$. The point ρ is a transverse multiple point of order N if in addition the gradient vectors are linearly independent.

8.3   Examples

We illustrate these definitions with a few examples.

8.3.1   Tandem Walks

Consider the class $\mathscr{T}$ of two-dimensional lattice walks with step set $\mathscr{S} = \{(1,0),(-1,1),(0,-1)\}$ restricted to the first quadrant. In Section 3.3.3, we express the generating function $T(z)$ that counts walks by length as a diagonal:

$$T(z) = \Delta \frac{\left((x^2 - y)(1 - \frac{1}{xy})(x - y^2)\right)}{(1 - z(y + xy^2x + x^2))(1 - x)(1 - y)} = \Delta \frac{G}{H_1 H_2 H_3}.$$

We denote $\mathbb{V}(H_j)$ by $\mathscr{V}_j$, $\mathbb{V}(H_j H_\ell)$ by $\mathscr{V}_{k\ell}$ and stratify $\mathscr{V}$. For each stratum we determine the critical points that lie in it. The results are summarized in Table 8.1. We use the notation $\omega = e^{2\pi i/3}$, a third root of unity. We visualize variety by considering $(\lvert x \rvert, \lvert y \rvert, \lvert z \rvert)$ for each $\mathscr{V}_i$ in

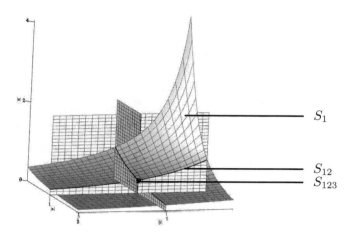

FIGURE 8.3

The image of $\mathcal{V}$ under the map $z \mapsto |z|$. It gives a general idea of the stratification, but it is misleading: $(w, w^2, w/3)$ is mapped to $(1, 1, 1/3)$ in the image, which is labelled $S_{1,2,3}$ but $(w, w^2/3, w/3)$ is in the stratum S_1.

Figure 8.3. In total there are three critical points:

$$(1, 1, 1/3), (w, w^2, w^2/3), (w^2, w, w/3).$$

Next we decide if they are transversal multiple points or not. Of the three, only the point $(1, 1, 1/3)$ is at the intersection of more than one variety, so the other two are not. One can further verify that $\mathcal{V}$ is actually smooth at the other two points. Intuitively, we can imagine from Figure 8.3 that if we move any one of the curves just a little bit, the three curves will continue to intersect in a single point. To formally test, we determine the matrix of gradients for H_1, H_2, H_3:

$$\begin{bmatrix} \nabla H_1(1,1,1/3) \\ \nabla H_2(1,1,1/3) \\ \nabla H_3(1,1,1/3) \end{bmatrix} = \begin{bmatrix} -1 & -1 & -3 \\ 1 & 0 & 0 \\ 0 & 1 & 0 \end{bmatrix}.$$

As this matrix of full rank, it confirms that $(1, 1, 1/3)$ is a transversal multiple point of order 3. We can conclude at this stage that the counting sequence has exponential growth $[z^n]T(z) = O(3^n)$.

8.3.2 Weighted Simple Walks

Weighted lattice walks provide an unique opportunity to examine how slight modifications in the geometry can affect the asymptotics. Consider a central weighting of the simple two-dimensional walks. In this model, the step set is $\mathscr{S} = \{(\pm 1, 0), (0, \pm 1)\}$, and a walk W ending at the point (k, ℓ) has weight $\phi(W) = a^k b^\ell$ for fixed positive real numbers a and b. A weighting that is independent of the path and depends only on the endpoint is a central weight. The quantity of interest is the polynomial, $q_{a,b}(n) = \sum_W \phi(W)$, where the sum is taken over all walks W of length n. How does this behave as n tends to infinity? How does it depend on a and b? These are questions we can answer.

Proposition 8.3. *The generating function for two-dimensional weighted simple walks satisfies:*

$$\sum_{n \geq 0} q_{a,b}(n) z^n = \Delta \left(\frac{G(x, y)}{H(x, y, z)} \right) \tag{8.1}$$

$$= \Delta \frac{a^{-1} b^{-1} (a^2 - x^2)(b^2 - y^2)}{1 - z(xy) S(a/x, b/y)} \cdot \frac{1}{(1 - x)(1 - y)}, \tag{8.2}$$

where $S(x, y)$ is the unweighted step inventory polynomial

$$S(x, y) = \sum_{\sigma \in \mathscr{S}} x^{\sigma_1} y^{\sigma_2}.$$

We identify $G(x, y)$ and $H(x, y, z)$ as the numerator and denominator of Equation (8.1) respectively. We label the factors of H as $H_1 = (1 - zxy S(a/x, b/y))$, $H_2 = (1 - x)$ and $H_3 = (1 - y)$. The manifold $\mathscr{V} \cap \partial D$ has several strata of interest, which we summarize. If (x, y) is in the closure of the domain of convergence, then $|x| \leq 1$ and $|y| \leq 1$.

The stratification of the intersection of $\mathscr{V}$ and the boundary of convergence is given in Table 8.2. The next step is to solve the critical equations on each of these varieties and to determine minimal critical points. This is summarized in Table 8.3.

The value of the height minimizer depends on the values of a and b, particularly their value relative to 1. For example, when they are

TABLE 8.2

Stratification of $\mathscr{V} \cap \partial D$

Stratum	Description
S_1	$\left\{ \left(x, y, (xyS(a/x, b/y))^{-1} \right) \mid x \neq 1, y \neq 1 \right\}$
S_{12}	$\left\{ \left(1, y, (yS(a, b/y))^{-1} \right) \mid y \neq 1 \right\}$
S_{13}	$\left\{ \left(x, 1, (xS(a/x, b))^{-1} \right) \mid x \neq 1 \right\}$
S_{123}	$\left\{ \left(1, 1, S(a, b)^{-1} \right) \right\}$

TABLE 8.3

Critical Points

Stratum	Critical point ρ	$\left\| \rho_1 \rho_2 \rho_3 \right\|^{-1}$	Caveat
S_1	$(a, b, \frac{1}{abS(1,1)})$	$S(1,1) = 4$	$a < 1, b < 1$
S_{12}	$(a, 1, \frac{1}{aS(1,b)})$	$S(1, b) = 2 + b + 1/b$	$a < 1$
S_{13}	$(1, b, \frac{1}{bS(a,1)})$	$S(a, 1) = a + 1/a + 2$	$b < 1$
S_{123}	$(1, 1, \frac{1}{S(a,b)})$	$S(a, b) = a + 1/a + b + 1/b$	

both less than 1, then the solution is given on S_1 with the minimal value of $S(a, b) = a + 1/a + b + 1/b$ across any real pair (a, b) is 4. The exponential growth is summarized in Table 8.4.

8.4 A Direct Formula for Powers

When the singular variety is smooth at a finitely minimal critical point, we have already seen a formula to determine the contribution

TABLE 8.4

Exponential Growth Factor of $q_{a,b}(n)$ Number of Weighted Simple Walks of Length n, with Weight Vector (a, b)

	$a < 1$	$a \geq 1$
$b < 1$	4	$a + 1/a + 2$
$b \geq 1$	$2 + b + 1/b$	$a + 1/a + b + 1/b$

at that point towards the dominant asymptotics. In such a case, there is little added difficulty to process the rational functions built by increasing the power of the denominator, as we shall soon see. In general, developing the formulas for the contribution of multiple points is beyond the scope of this text, but we now have sufficient vocabulary to parse the statements of the formulas. In each case, as before, the complete asymptotic estimate comes from a sum of contributions:

$$[\mathbf{z}^{n\mathbf{r}}]\frac{G(\mathbf{z})}{H(\mathbf{z})} \sim \sum_{\rho \in \text{Contributing}} \Phi_\rho(n) \quad \text{as } n \to \infty. \tag{8.3}$$

We concentrate on results to compute $\Phi_\rho(n)$, the contribution towards $[\mathbf{z}^{n\mathbf{r}}]F(\mathbf{z})$. Precise estimates on the error are possible to compute.

Theorem 8.4. *Suppose* $F(\mathbf{z}) = \frac{G(\mathbf{z})}{H(\mathbf{z})^\ell}$ *and* ρ *minimizes the height function at* $\mathbf{r}$. *Furthermore suppose that* ρ *is a multiple point of order* k, *and that* G *is analytic and non-zero at* ρ. *The value of* $\Phi_\rho(n)$ *is*

$$\Phi_\rho(n) \sim (2\pi)^{(1-d)/2}\binom{-nr_k}{\ell - 1}(\det\mathcal{H})^{-1/2}\frac{G(\rho)}{\rho_k\frac{\partial H}{\partial z_k}(\rho)}. \tag{8.4}$$

This expression is combinatorial, and satisfyingly compact. It allows us to see the impact of the exponent of H in the formula.

Example 8.2 (Generalized binomial). Let us recover a known quantity: Set $F(z) = \frac{1}{(1-z)^\ell}$ and $\mathbf{r} = 1$. Then $\rho = 1$, and we set $k = 1$, since the variety $\mathcal{V} = \{1\}$ is zero dimensional (and hence it is the stratum containing 1). By Theorem 8.4,

$$[z^n]F(z) \sim \binom{-n}{\ell - 1} = (-1)^{n-\ell-1}\binom{n+\ell-2}{\ell - 1},$$

which we recognize as the generalized binomial, correct up to a sign.

◄

8.5 The Contribution of a Transverse Multiple Point

Next we consider the more general problem. Roughly, if the pole is order k, then the inner residue integrals are over a $(d-k)$ dimensional surface. Thus we can estimate the contribution of ρ in direction $\mathbf{r}$ will be approximately $\Phi_\rho(n) \sim C\left(\rho_1^{r_1} \cdots \rho_d^{r_d}\right)^{-n} n^{(d-k)/2}$, (for some constant C) once we incorporate the drop in dimension.

Suppose we know that ρ is a multiple point, but that we do not have the tidy factorization. It may be possible to compute directly from H. This can be done explicitly when $d = k = 2$.

Theorem 8.5. *Let* $F(x,y) = \frac{G(x,y)}{H(x,y)}$. *Suppose that H factors into two distinct, square-free factors, $H = H_1 H_2$ and that the varieties $\mathcal{V}_{H_1}$ and $\mathcal{V}_{H_2}$ have a transversal intersection at ρ a contributing minimal critical point minimizing the height function at $\mathbf{r}$.*

If G is holomorphic in a neighbourhood of ρ and $G(\rho) \neq 0$, then

$$\Phi_\rho(n) = \left(\rho_1^{-r_1} \rho_2^{-r_2}\right)^n \frac{G(\rho)}{\sqrt{H_{xy}^2(\rho) - H_{xx}(\rho)H_{yy}(\rho)}}. \tag{8.5}$$

This is a specialization of a more general result to handle transverse multiple points whose order matches the dimension.

Theorem 8.6. *Suppose that $F(\mathbf{z}) = G(\mathbf{z})/\prod_{j=1}^k H_j(\mathbf{z})^{m_j}$ and ρ is a transversal multiple critical point for $\mathbf{r}$. Suppose the stratum containing ρ is zero dimensional. Then*

$$\Phi_\rho(n) = \frac{1}{(m-1)!} \frac{G(\rho)}{\det \Gamma_\Psi(\rho)} \left((n\mathbf{r})\Gamma_\Psi^{-1}\right)^{m-1}.$$

We can unpack this statement a little bit. First, suppose the singular variety $\mathcal{V}$ locally decomposed around ρ as $\cup_{j=1}^k \mathcal{V}_j$ then $S = \cap_{j=1}^k \mathcal{V}_j$. When $k = d$, the augmented matrix of logarithmic gradients is

$$\Gamma_\Psi(\mathbf{z}) := \left[z_\ell \frac{\partial H_j(\mathbf{z})}{\partial z_\ell}\right]_{\ell,j}.$$

We recall the notation used:

$$\frac{1}{(m-1)!} = \frac{1}{\prod(m_i - 1)!} \qquad x^{m-1} = x_1^{m_1-1} \cdots x_d^{m_d-1}.$$

Note, G is not necessarily a polynomial. It could be meromorphic, but it is well-behaved around the critical points. This allows us to group factors.

Example 8.3. We continue the example of the tandem walks, and focus on the contribution of the multiple point, $\rho = (1, 1, 1/3)$. We notice that $G(1,1) = 0$, and thus we cannot apply the theorem as we do not satisfy the hypotheses! What we can do in this case is reduce one of the factors in the numerator. In this case, we rewrite the rational using a reduction: $x^2 - y = (x-1)(x+1) - (y-1)$:

$$\frac{G}{H} = \frac{x+1}{(1-x)(1-z(x+y^2+x^2y))} + \frac{1}{(1-y)(1-z(x+y^2+x^2y))} \qquad (8.6)$$

For each term in the sum the numerator is not cancelled. We apply the theorem to each part to deduce the contribution of ρ:

$$\Phi_{(1,1,1/3)}(1,1,1) = 3^n \cdot n^{-3/2} \cdot \frac{3\sqrt{3}}{4\sqrt{\pi}} + O(3^n \cdot n^{-5/2}).$$

The two smooth points give a contribution of smaller polynomial order, $O(3^n/n^2)$, which we can compute using the results from the previous chapter.   ◄

In the case that $k < d$, the computation is a little more subtle. There is an explicit formula in Section 10.3.2 of [PW13]. Roughly, the order of the multiple point is incorporated into the formula using the **augmented log normal** matrix. The matrix Γ_ψ is defined as follows. The first k rows are the logarithmic gradients:

$$z_\ell \partial_\ell H_j,$$

as before. However, the remaining entries make a square matrix: The last $d - k$ rows are the vectors $\rho_{\pi_i} e_{\pi_i}$, $i = 1 \ldots, d - k$ are chosen so that the matrix is nonsingular. There may be more than one valid choice, but the result should be the same. The formulas to complete this case are given in and rely on the logarithmic parametrization of H.

8.6 Discussion

Pemantle and Wilson developed the criteria for fixed points. Proposition 8.1 is discussed in [PW13, Section 8.3]. Furthermore, Theorem 8.6 appears as [PW13, Theorem 10.3.3].

A more precise description of the criteria for minimal points is given in Proposition 10.3.6 of [PW13]. It should be noted that as stated, the formulas are correct only up to a sign. Formulas that determine the sign require more detailed information about the curvature of the variety. This can be made somewhat explicit, but it takes much more work.

It is possible to replicate the enumeration results for lattice walks in any dimension. Of course, weights of models can be thought of as probabilities. There are several examples of this kind of analysis in the literature. Pemantle [BGPP10], Courtiel, Melczer, Mishna and Raschel [CMMR17], Melczer and Wilson [MW18], and Mishna and Simon [MS19]. That said, analysis of the more general problem of reflectable walks in Weyl chambers appears much earlier as noted in the Discussion of the previous chapter.

8.7 Problems

Exercise 8.1. Using the formal definition of multiple point, prove that crossing lines intersect transversely in two dimensions, but not in three. $\quad\square$

Exercise 8.2 (Tandem walks of free endpoint). Do the analysis for the tandem walks that end anywhere. Hint, the denominator is not a transversal intersection.

Reference: [MW18]

$\square$

Exercise 8.3 (Double tandem). Let $\mathscr{S} = \{\leftarrow, \nwarrow, \uparrow, \rightarrow, \searrow, \downarrow\}$. Determine the asymptotic number of excursions.

Reference: [BMM10]

$\square$

Exercise 8.4 (The simple walks). Consider the three-dimensional simple walks, where the step set is the set of elementary vectors and their negatives:
$$\mathscr{S} = \{\pm(1,0,0), \pm(0,1,0), \pm(0,0,1)\}.$$

The following integer weighting of the steps is central:

Step	$(1,0,0)$	$(-1,0,0)$	$(0,1,0)$	$(0,1,0)$	$(0,0,1)$	$(0,0,1)$
Weight	8	2	4	4	1	16

Determine a, b, c, d so that the weight of each step (i, j, k) is $d\, a^i b^j c^k$ to show that the model is centrally weighted. Determine the asymptotic number of walks of length n.   ❏

9

Partitions

CONTENTS

9.1	Integer Partitions	192
9.2	Vector Partitions	194
	9.2.1 Integer Points in Polytopes	195
9.3	Asymptotic Analysis	197
	9.3.1 The Singular Variety and Hyperplane Arrangements ..	197
	9.3.2 Reducing to the Case of Transversal Intersection	199
	9.3.3 Algebraic Independence	200
	9.3.4 Decomposition Dictionary	202
	9.3.5 Decomposition into Circuit-free Denominators	202
	9.3.6 The Complete Reduction Algorithm	203
	9.3.7 How to Compute a Reduction Rule	204
9.4	Asymptotics Theorem	204
9.5	An Example	205
	9.5.1 An Exact Solution	206
	9.5.2 An Asymptotic Solution	206
	9.5.3 The Bases with No Broken Circuits	207
	9.5.4 The Reduction Algorithm	207
	9.5.5 Asymptotic Formula	209
9.6	Discussion	210
9.7	Problems	210

The behaviour of the coefficients of a series expansion of a multivariable rational function is driven by the geometry of the singular variety at the contributing critical points. In the simplest case, the variety is smooth at the contributing critical points. If the variety locally decomposes into a union of smooth varieties and the intersection is transversal, this is more complicated but manageable, as we saw in the last chapter. When the denominator factors into powers, say H^2, the variety is the same, but the asymptotic behaviour is slightly modified. Next we consider the case of critical points that are not transversal multiple points. We do not consider this case in full generality. Rather, we consider a situation that arises in combinatorics, where the singular variety under the relog map is essentially a **hyperplane arrangement**. Hence, these are known as **arrangement points**. We leverage the underlying combinatorics of hyperplane arrangements to reduce the problem to the case of transverse multiple points. The reduction process is the central topic of this chapter.

We have seen how to compute the contribution from multiple points at a transversal intersection. When this is not the case, we can try to rewrite the function as a sum of rationals with fewer factors in the denominator. It turns out that there is a multivariable version of the partial fraction decomposition process to decompose a rational. We can illustrate how to exploit a particular algebraic dependence defining the hyperplane arrangements underlying the singular variety and describe to give a terminating algorithm.

One large family of combinatorial classes that arise in this context are the **vector partitions**. Their generating functions are used to determine polytope point enumerators, which appear in a very wide range of families of combinatorial problems from representation theory to discrete optimization. We start by defining these objects and then consider how to determine asymptotic enumeration formulas.

9.1   Integer Partitions

One of the most fundamental combinatorial classes is the set of integer partitions, denoted $\mathscr{P}$. A **partition** of $n \in \mathbb{N}$ is an unordered set of positive integers that sum to n. They are typically represented by a decreasing sequence λ and in that case we write $\lambda \vdash n$. For example, the seven partitions of size 5 are

$$\mathscr{P}_5 = \{(5), (4,1), (3,2), (3,1,1), (2,2,1), (2,1,1,1), (1,1,1,1,1)\}.$$

Each summand is a **part** of the partition. The number of parts of a partition is its **length**. A common representation of a partition is by its **Ferrers diagram**, which is a stacked collection of rows of •. For example, the Ferrers diagram of the partition $(7, 4, 2, 2, 1)$ is represented

As a combinatorial object, we view a partition as a sequences of rows in the Ferrers diagram, and each row is a sequence of atoms. The size is the total number of atoms. Suppose that the parts are restricted to some finite set $\mathscr{S}$. The classic example considers parts whose sizes resemble coin denominations, for example $\mathscr{S} = \{1, 5, 10, 25, 100\}$. The number of partitions of n with parts in $\mathscr{S}$ is precisely the number of ways to express n cents using coins with denominations 1, 5, 10, 25 or 100. We represent each part of value k by the class with a single element of size k, $\text{SEQ}_{=k}(\bullet)$, and hence the parts of length k in a partition form a set of these pieces, which we represent by ordering the elements. The class $\mathscr{P}^{\mathscr{S}}$ of partitions of integers using only parts with sizes from $\mathscr{S}$ is S-regular, and satisfies

$$\mathscr{P}^{\mathscr{S}} = \prod_{s \in \mathscr{S}} \text{SEQ}(\text{SEQ}_s \bullet).$$

Consequently, the generating function satisfies

$$P^{\mathscr{S}}(x) = \prod_{s \in \mathscr{S}} \frac{1}{1 - x^s}.$$

This generating function has radius of convergence of 1 with a pole at 1 of order $|\mathscr{S}|$, and we can deduce immediately that the number of partitions with parts from $\mathscr{S}$ is $O(n^{|S|-1})$. When $\mathscr{S}$ is finite, the generating function is a simple $\mathbb{N}$-rational function that we can analyze.

Managing the periodicity is a significant aspect of partition enumeration. Indeed, it is difficult to determine even the largest value n, such that $[x^n]P^{\mathscr{S}}(x) = 0$.

Example 9.1. The number of ways to express 21 cents as a sum of coins in denomination $\{1, 5, 10, 25, 100\}$ is precisely

$$[x^{21}]\frac{1}{1 - x}\frac{1}{1 - x^5}\frac{1}{1 - x^{10}}\frac{1}{1 - x^{25}}\frac{1}{1 - x^{100}} = 9.$$

For n large, the number of ways to express n cents is $O(n^4)$. ◄

What is less obvious, and requires justification, is that such a formula is true also for infinite sets $\mathcal{S}$ of part sizes. For example, we could define the class of all partitions as a set limit

$$\mathcal{P} := \lim_{N \to \infty} \prod_{k=1}^{N} \text{SEQ}(\text{SEQ}_k \bullet), \qquad (9.1)$$

and deduce that the generating function for integer partitions $\mathcal{P}$ is

$$P(x) := \sum_{\substack{\lambda \in \mathcal{P} \\ \lambda \vdash n}} x^n = \prod_{k \geq 1} \frac{1}{1 - x^k}. \qquad (9.2)$$

Given the form, we see immediately that $P(x)$ has multiple singularities at 1 and indeed at all integral roots of unity. Since $P(x)$ has an infinite number of singularities, it is not D-finite. The asymptotics of the partition number is subtle.

9.2   Vector Partitions

A **vector partition**, or when the context is clear simply a partition of $\mathbf{r} \in \mathbb{N}^d$, is a set of vectors in $\mathbb{N}^d \setminus \{0\}$ whose (component-wise) sum is equal to $\mathbf{r}$. When appropriate, we retain the same notation and vocabulary as the univariate case. For example,

$$\lambda = \left\{ \begin{bmatrix} 3 \\ 2 \end{bmatrix}, \begin{bmatrix} 1 \\ 2 \end{bmatrix}, \begin{bmatrix} 1 \\ 1 \end{bmatrix}, \begin{bmatrix} 1 \\ 1 \end{bmatrix} \right\} \vdash \begin{bmatrix} 6 \\ 6 \end{bmatrix}.$$

Let $\mathcal{P}$ denote a family of d-dimensional vector partitions. The multivariable generating function for $\mathcal{P}$ where x_i marks the value of the ith component is

$$P(\mathbf{x}) := \sum_{\lambda \in \mathcal{P}} \sum_{\lambda \vdash \mathbf{r}} x_1^{r_1} \dots x_d^{r_d} = \sum_{\mathbf{r} \in \mathbb{N}^d} p^{\mathcal{P}}(\mathbf{r}) \mathbf{x}^{\mathbf{r}}. \qquad (9.3)$$

Suppose that the parts come from a finite set $\mathcal{S}$. We model this with building blocks built from d types of atoms, each one marked by its own variable:

$$\mathcal{P}^{\mathcal{S}} \equiv \prod_{s \in \mathcal{S}} \text{SEQ} \begin{bmatrix} \text{SEQ}_{s_1} \, ① \\ \vdots \\ \text{SEQ}_{s_d} \, ⓓ \end{bmatrix}.$$

We conclude that

$$P^{\mathscr{S}}(\mathbf{x}) = \prod_{s \in \mathscr{S}} \frac{1}{1 - x_1^{s_1} \dots x_d^{s_d}}. \tag{9.4}$$

Thus, the number of vector partitions of $\mathbf{r}$ with parts in $\mathscr{S}$ is

$$p^{\mathscr{S}}(\mathbf{r}) = [\mathbf{x}^{\mathbf{r}}] \prod_{s \in \mathscr{S}} \frac{1}{1 - x_1^{s_1} \dots x_d^{s_d}}.$$

Example 9.2. Let $\mathscr{S} = \left\{ \begin{bmatrix} 1 \\ 1 \end{bmatrix}, \begin{bmatrix} 1 \\ 2 \end{bmatrix}, \begin{bmatrix} 2 \\ 3 \end{bmatrix} \right\}$. The number of partitions of $\begin{bmatrix} 6 \\ 6 \end{bmatrix}$ is

$$[x^6 y^6] \frac{1}{1 - xy} \frac{1}{1 - xy^2} \frac{1}{1 - x^2 y^3} = 2.$$

We can list them explicitly (using a slightly more compact notation):

$$\begin{bmatrix} 1 \\ 1 \end{bmatrix}^6, \quad \begin{bmatrix} 3 \\ 2 \end{bmatrix} \begin{bmatrix} 1 \\ 2 \end{bmatrix} \begin{bmatrix} 1 \\ 1 \end{bmatrix}^2.$$

◄

Suppose the possible parts of a vector partition are given by the multi-set $\mathscr{S} = \{s_1 \dots s_N\}$. A partition of $\mathbf{r}$ using these parts is specified by an N-dimensional vector $\mathbf{m} \in \mathbb{N}^N$

$$m_1 s_1 + \dots + m_N s_N = \mathbf{r}. \tag{9.5}$$

Note, there is a partition of $\mathbf{r}$ with parts in $\mathscr{S}$ if there is a solution $\mathbf{x} = \mathbf{m} \in \mathbb{N}^N$ to the linear algebra system

$$A\mathbf{x} = \mathbf{r},$$

where A is the $d \times N$ matrix

$$A = \begin{bmatrix} | & & | \\ s_1 & \dots & s_N \\ | & & | \end{bmatrix}.$$

9.2.1 Integer Points in Polytopes

Polytopes provide a geometric interpretation of a set of vector partitions. A **polytope** is a set of points defined as either the convex hull of a set of points, or the (bounded) intersection of half-spaces defined by hyperplanes:

$$\mathbb{P} = \{\mathbf{x} \in \mathbb{R}^d \mid \mathbf{a}_1 \cdot \mathbf{x} \le r_1, \cdots, \mathbf{a}_d \cdot \mathbf{x} \le r_d\} \tag{9.6}$$

$$= \{\mathbf{x} \in \mathbb{R}^d \mid A\mathbf{x} \le \mathbf{r}\} \tag{9.7}$$

where

$$A = \begin{bmatrix} -- & \mathbf{a}_1 & -- \\ & \vdots & \\ -- & \mathbf{a}_d & -- \end{bmatrix}.$$

Consider $\mathbb{P} \cap \mathbb{N}^d$. This is a set of integer lattice points inside the polytope, the **discrete volume** of the polytope. If we add slack variables to convert the inequalities into equalities, then we see immediately that each point $\mathbf{m}$ inside $\mathbb{P} \cap \mathbb{N}^d$ corresponds to a vector partition of $\mathbf{r}$. Let A' be this new matrix.

The n-th dilate of the polytope $\mathbb{P}$ is

$$n\mathbb{P} = \{(nx_1, \ldots, nx_d) \mid \mathbf{x} \in \mathbb{P}\} \tag{9.8}$$

$$= \{n\mathbf{x} \in \mathbb{R}^d \mid A'\mathbf{x} = \mathbf{r}\} \qquad = \{\mathbf{x} \in \mathbb{R}^d \mid A'\mathbf{x} = n\mathbf{r}\}, \tag{9.9}$$

since we expand by a scalar factor.

The **lattice point enumerator for the nth dilate** is the sequence $L_{\mathbb{P}}(n) := |\mathbb{Z}^d \cap n\mathbb{P}|$. The sequence $L_{\mathbb{P}}(n)$ has many interesting, combinatorial properties.

Theorem 9.1 (Erhart theorem). *Suppose that $\mathbb{P}$ is an integral convex d-dimensional polytope, then $L_{\mathbb{P}}(n)$ is a quasi-polynomial in n of degree d. The constant term is 1 and the volume of $\mathbb{P}$ is $[n^d]L_{\mathbb{P}}(n)$.*

Example 9.3. The **standard d-dimensional simplex** is the polytope[1]

$$\mathbb{P}_{d-1} = \{\mathbf{x} \in \mathbb{R}^d \mid x_1 + \cdots + x_d = 1; \quad x_i \ge 0, i = 1, \ldots, d\}.$$

The n-th dilate of this polytope is thus

$$n\mathbb{P}_{d-1} = \{\mathbf{x} \in \mathbb{R}^d \mid x_1 + \cdots + x_d = n; \quad x_j \ge 0, j = 1, \ldots, d\}.$$

---

[1] Typically, the simplex polytope is denoted by the symbol Δ, but the notation clashes here.

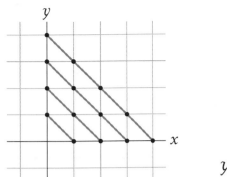 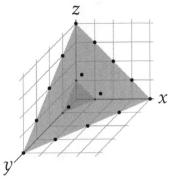

FIGURE 9.1
Simplex polytopes of low dimension and some dilates: *(left)* $n\mathbb{P}_1$, $n = 1,\ldots,4$; *(right)* $n\mathbb{P}_2$, $n = 1,4$.

The lattice point enumerator of $\mathbb{P}_{d-1}$ is

$$L_{\mathbb{P}_{d-1}}(n) = \binom{d+n-1}{d-1} = (n+d-1)(n+d-2)\ldots(n+1)/d! = \frac{1}{2}n^{d-1} + \cdots + 1.$$

◄

Example 9.4. We see in Figure 9.1 (right) $\mathbb{P}_{3-1}(4) = 15$ which is precisely $\binom{2+4}{2}$. The leading term is $1/2$, which is the area of the triangle with endpoints e_1, e_2, e_3.

◄

9.3 Asymptotic Analysis

Rational functions of the form in Eq. (9.10) typically have more factors in the denominator than the dimension, and thus a priori we know that at any critical point the intersection cannot be transversal. However, the structure has some simplifying characteristics: the numerator is always 1, and each factor in the denominator is of the same form of 1 minus a monomial. This section presents a strategy that will decompose the sum into components that can be analyzed with the methods of the previous chapter. The proofs of correctness, and the handling of degenerate cases is beyond the scope of this text, but can be found in Chapter 10 of [PW13].

Problem Statement

Let $A = [a_{ij}]$ be an $d \times N$ matrix with entries in $\mathbb{N}$, and let $\mathbf{r} \in \mathbb{N}^d$. Determine exact or asymptotic information about the quantity $[x^{n\mathbf{r}}]F(\mathbf{x})$ for

$$F(\mathbf{x}) = \prod_{k=1}^{N} \frac{1}{1 - x_1^{a_{i1}} \cdots x_d^{a_{id}}}. \tag{9.10}$$

9.3.1   The Singular Variety and Hyperplane Arrangements

Let

$$H(\mathbf{x}) = H_1(\mathbf{x}) \cdots H_N(\mathbf{x}) = \prod_{j=1}^{N} \left(1 - x_1^{a_{j1}} \cdots x_d^{a_{jd}}\right). \tag{9.11}$$

To start, we examine the singular variety, $\mathscr{V} = \{\mathbf{x} \mid H(\mathbf{x}) = 0\}$. It is a union of the varieties $\mathscr{V}_{H_j}$ for $j = 1, \ldots, N$. Points on the subvariety $\mathscr{V}_{H_j}$ satisfy $1 = x_1^{a_{j1}} \ldots x_d^{a_{jd}}$. Equivalently,

$$0 = -a_{j1} \log |x_1| \cdots - a_{jd} \log |x_d|.$$

That is, the image of the points in the variety under the relog map is a collection of hyperplanes in $\mathbb{R}^d$. We see in Figure 9.2 an example in dimension 3. The left-hand side of the figure presents a representation of the positive real elements of the variety. The right-hand side is the image of the three curves in the relog map: a collection of three hyperplanes intersecting at the origin. We are able to leverage these connections to rewrite the rational function in a form amenable to analysis.

Indeed, this phenomenon is general. Under some nondegeneracy conditions that we return to in a moment, we see that in any dimension, the origin is a point in $\{(-\log |x_1|, \ldots, -\log |x_d|) \mid \mathbf{x} \in \mathscr{V}_H\}$ for H of the form Eq. (9.10). Consequently, the origin minimizes the height function **for any choice of r**, and the all-ones vector denoted $\mathbf{1}$ is a point that minimizes $\left| x_1^{r_1} x_2^{r_2} \ldots x_d^{r_d} \right|^{-1}$. There may be other points on the torus of $\mathbf{1}$ that satisfy the critical point equations, but because the series is combinatorial, we will always have $\mathbf{1}$ as a critical point.

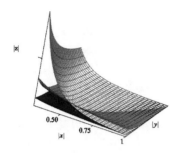
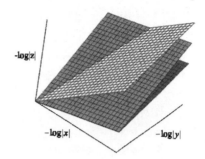

FIGURE 9.2
Three intersecting varieties: *(left)* The real points of $\mathscr{V}_{1-xyz}$, $\mathscr{V}_{1-xy^2z}$, and $\mathscr{V}_{1-x^3yz}$. Remark that all three intersect at $(1,1,1)$. *(right)* The image of the three varieties under -relog. The image of each variety is a hyperplane which passes through the origin.

The Critical Point

The unique real, positive critical point of the function

$$F(\mathbf{x}) = \frac{1}{H(\mathbf{x})} = \prod_{k=0}^{N} \frac{1}{1 - x_1^{a_{i1}} \cdots x_d^{a_{id}}}, \qquad (9.12)$$

in the direction $\mathbf{r} \in \mathbb{N}^d$ is the point $\mathbf{1} = (1,1,\dots,1)$. Consequently, the exponential growth of any polytope point enumerator is 1.

If $H(\mathbf{x})$ factors, i.e., $N > 1$, then $\mathbf{1}$ is a multiple critical point. If $H(\mathbf{x})$ has more factors than the dimension, $N > d$, then the variety does not have a transversal intersection at $\mathbf{1}$. There may be other critical points on the torus of $\mathbf{1}$.

9.3.2 Reducing to the Case of Transversal Intersection

Recall in the univariate rational case the first step is to use partial fraction decomposition to decompose a rational into a sum of terms, each with a linear factor, or a power of a linear factor. The equivalent in d dimensions is the following result of Leinartas.

Proposition 9.2. *Every rational expression in d variables can be represented as a sum of rational expressions each of whose denominators contains at most d unique irreducible factors.*

The proof of this result is constructive, meaning that we can determine the decomposition algorithmically. There are two main steps. The first uses a Nullstellensatz decomposition, and the second uses algebraic independence. The latter is better phrased in terms of the matrix, which defines the problem, as it corresponds to linear independence of columns. Each summand will have a transverse intersection at **1**. Recall, a point is a transverse multiple point of order n if and only if in addition to being a critical point, the gradient vectors $\{\nabla H_j(\mathbf{1}) \mid 1 \leq j \leq n\}$ are linearly independent. In that case, we can apply Theorem 8.6 summarized below to determine the contribution from each summand.

The asymptotic formula for the coefficients along a fixed direction is given by a sum of the contributing formulas, namely if $\mathbf{r}$ is in the support of the summand. In general there is not a single formula for the polytope point enumerator – it will depend on $\mathbf{r}$.

Let us be more precise about the result central to this case.

Theorem 9.3. *Let $F = \frac{G}{H}$ with $G, H \in K[x_1, \ldots, x_d]$. Let $H = H_1^{m_1} \ldots H_N^{m_N}$ be the unique factorization of H in $K[x_1, \ldots, x_d]$, and let $\mathcal{V}_k = \{\mathbf{x} \in \overline{K}^d \mid H_k(\mathbf{x}) = 0\}$, the algebraic variety of H_i over $\overline{K}$. Then $F(\mathbf{x})$ satisfies*

$$F(\mathbf{x}) = \sum_J \frac{G_J(\mathbf{x})}{\prod_{j \in J} H_j^{b_{j,J}}(\mathbf{x})}, \qquad (9.13)$$

where the $b_{j,J}$ are positive integers (possibly bigger than the m_j), the G_J are polynomials in $K[\mathbf{x}]$ (possibly 0). The sum is taken over subsets $J \subset \{1, \ldots, m\}$ such that $\cap_{j \in J} \mathcal{V}_j \neq \emptyset$, and $\{H_j \mid j \in J\}$ is algebraically independent.

This result is made effective in two stages:

Determine the index sets J In the first stage, we determine the sets J, which ensure that $\{H_j \mid j \in J\}$ is algebraically independent. We use vocabulary and concepts from matroid theory to facilitate the exposition and to justify termination of the process.

Decompose $F(\mathbf{x})$ We use the index sets and Nullstellensatz decomposition to write $F(\mathbf{x})$ in the form Eq. (9.13). Some further tidying and local simplifications are also done at this point.

9.3.3 Algebraic Independence

In order to ensure the required algebraic independence, we use the linear algebra structure central to the problem. Matroid theory is an abstraction of common notions of dependence across mathematics. It is defined in terms of a ground set E and a family $\mathcal{B}$ of subsets of E. A set $B \in \mathcal{B}$ is a **basis** and they individually and collectively satisfy a number of important properties. The term is taken from linear algebra, and this is the right intuition. A basis B is independent, and all of its subsets are independent. For any choice of $e \in E \setminus B$, the set $B \cup \{e\}$ is dependent. A **circuit of a matroid** is a dependent set minimal in that all of its proper subsets are all independent.

For matrix A, the associated matroid $M(A)$ is defined with respect to the ground set that is the column labels of A. A subset of E is independent if the associated column vectors are linearly independent, and dependent otherwise. We write $\mathcal{B}$ to denote the set of bases.

The first part of the reduction algorithm establishes the set of indices J in Eq. (9.13). Let the set of valid indices be denoted $\mathcal{J}$.

1. Label the columns of A from 1 to N. Set $E = \{1 \ldots, N\}$;

2. Determine the circuits C of E. These are sets of linearly dependent columns but removing any column breaks the dependancy;

3. For each circuit c construct a "broken circuit," which is c minus its largest element.

4. We set $\mathcal{J}$ to be those bases that *do not* contain any broken circuit.

Now, not every element J of $\mathcal{J}$ will contribute a term in Eq. (9.13) such that $G_J \neq 0$, but each summand will contribute a term that we can analyze.

Example 9.5. Let A be the matrix

$$A = \begin{bmatrix} 1 & 0 & 1 & 1 & 3 \\ 0 & 1 & 1 & 2 & 1 \\ 0 & 0 & 1 & 1 & 1 \end{bmatrix}.$$

The ground set of $M(A)$ is $E = \{1, 2, 3, 4, 5\}$. Note that the set $\{1, 3, 5\}$ (which we shorten to 135) corresponds to the submatrix

$$\begin{bmatrix} 1 & 1 & 3 \\ 0 & 1 & 1 \\ 0 & 1 & 1 \end{bmatrix},$$

which has rank 2, hence 135 is not an independent set in $M(A)$. Any pair of columns is independent, so 135 is a circuit. Similarly, 234 is a circuit. The remaining sets of size 3 are bases:

$$\mathcal{B} = \{123, 124, 125, 134, 145, 235, 245, 345\}.$$

The circuits are the subsets of E of size four, except those that contain 135 or 234. The circuits and broken circuits of $M(A)$ are respectively $C = \{1245, 135, 234\}$ and $BC = \{124, 13, 23\}$. The bases which do not contain a broken circuit are

$$\mathcal{F} = \{125, 145, 245, 345\}.$$

◄

Why does this work? For this problem, *the matrix A is equal to the matrix of gradients* Φ *for the* H_j, *evaluated at* **1**. Thus, by definition, the columns given by bases are linearly independent, as will be the logarithmic gradients of their corresponding factors. This ensures a transversal intersection at **1** for each summand.

9.3.4   Decomposition Dictionary

Next we show that given a circuit, we can determine a polynomial equation between the corresponding polynomials.

Lemma 9.4 (Dictionary). *Let c be any circuit of $M(A)$. There is a collection* $\{P_j(x) : j \in c\}$ *of invertible elements of* $K[[x]]$, *such that* $\sum_{j \in c} P_j(x) H_j(x) = 0$.

We can set up a dictionary to reduce broken circuits. For every circuit, we write the factor-indexed largest element in the circuit in terms of the remaining elements: $H_\ell = \sum_{j \in c \setminus \ell} P_j(\mathbf{x}) H_j(\mathbf{x})$.

9.3.5   Decomposition into Circuit-free Denominators

The reduction manipulation is very simple. Let us illustrate it on a simple circuit, say $c = 1245$ from the example above. Because this is a circuit we can find a rewriting rule: $H_5 = P_1 H_1 + P_2 H_2 + P_4 H_4$.

$$\frac{1}{H_1 H_2 H_4 H_5} = \frac{1}{H_1 H_2 H_4 H_5} \cdot \frac{H_5}{H_5} \tag{9.14}$$

$$= \frac{P_1 H_1 + P_2 H_2 + P_4 H_4}{H_1 H_2 H_4 H_5^2} \tag{9.15}$$

$$= \frac{P_1}{H_2 H_4 H_5^2} + \frac{P_2}{H_1 H_4 H_5^2} + \frac{P_4}{H_1 H_2 H_5^2}. \tag{9.16}$$

Note that the initial quotient contained a circuit, and after the reduction we are left with a sum of three quotients. The indices of the factors are, respectively, $245, 145, 125$. Note that these are all bases which do not contain a broken circuit.

The algorithm proceeds by iteratively applying this reduction. Roughly speaking, it terminates because at each stage we reduce the size of the index set of the factors.

9.3.6   The Complete Reduction Algorithm

We start with $F(\mathbf{x}) = 1/H(\mathbf{x})$ of the form Equation 9.11, and first compute $\mathcal{J}$ = set of bases of $M(A)$, none of which contain a broken circuit. We build in expression of $F(\mathbf{x})$ of the form Eq. 9.13 such that the numerators are non-vanishing at $\mathbf{1}$.

1. Set the current expression to be $\frac{1}{H(\mathbf{x})}$, with a single term.

2. If any term in the current decomposition has a denominator whose terms contain a circuit, c, apply the reduction step above using this circuit. Repeat until no longer possible.

3. Collect terms with the same denominator.

4. Maximally reduce the rational expressions. For each term $G_J(\mathbf{x})/H_J(\mathbf{x})$ in the current expression, check whether the numerator contains terms in the ideal generated by the denominator. If there are such terms, choose one among them whose denominator has maximum degree and replace it by a sum of terms with smaller support.

5. For each quotient $G_J(\mathbf{x})/H_J(\mathbf{x})$ in the expression ensure that $G_J(1) \neq 0$. If it does, rewrite the quotient into a sum with the same factors but with smaller powers. Again, repeat until no longer possible.

In general, one cannot reduce further than this.

9.3.7 How to Compute a Reduction Rule

Given a circuit, we find $p_j(\mathbf{x}) \in K[\mathbf{x}]$, such that

$$0 = \sum_{j \in \mathscr{J}} p_j(\mathbf{x})H_j(\mathbf{x}).$$

This allows us to write $H_{j*} = \sum_{\substack{j \neq j* \\ j \in \mathscr{J}}} q_j(\mathbf{x})H_j(\mathbf{x})$ for q_j.

If the problem is small enough, we may find a solution p_j by hand by substituting small formal polynomials, expanding and solving. Ideally, we find a solution with $p_{j*} \in K$. For larger, and certainly in the case of the more interesting, problems, one should use a computer algebra program.

9.4 Asymptotics Theorem

We repeat the theorem to deduce the asymptotics as a measure of convenience.

Theorem 9.5. *Suppose*

$$F(\mathbf{x}) = \frac{G(\mathbf{x})}{\prod_{j=1}^{d} H_j(\mathbf{x})^{m_j}}$$

has a series expansion $\sum f(\mathbf{n})\mathbf{x}^\mathbf{n}$ in a non-empty domain of convergence containing the origin. Assume that each H_j is square-free and all divisors intersect transversally at a unique contributing critical point $\rho \in \mathbb{C}^d$ and that G, holomorphic in a neighbourhood of ρ, satisfies $G(\rho) \neq 0$. Then

$$f(\mathbf{r}) \sim \frac{1}{\prod(m_i - 1)!} \frac{G(\rho)}{\det \Psi} \rho^\mathbf{r}(\mathbf{r}\Psi^{-1})^{(m_1-1,\ldots,m_d-1)}, \tag{9.17}$$

where Ψ is the $d \times d$ matrix $\left[\rho_\ell \frac{\partial H_k}{\partial x_\ell}(\rho)\right]_{0 \leq k, \ell \leq n}$.

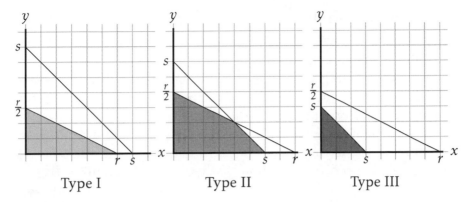

Type I   Type II   Type III

FIGURE 9.3
Three possibilities for the polytope defined by the inequalities $x + y = r$
and $2x + y = s$.

9.5 An Example

We illustrate the process on an example. It is small enough to determine the lattice enumerator explicitly, but it is complex enough to
illustrate the general process for treating hyperplanes. It is also de-
signed to have a single critical point at $(1, 1)$ in order to simplify the
process. We determine the enumerative information on the number
of vector partitions of (r, s) with parts in $\mathscr{S} = \{(1, 0), (0, 1), (1, 1), (1, 2)\}$.
Equivalently, this is the lattice point enumerator for the two-
dimensional polytope in the first quadrant bound by the hyperplanes:

$$x + y = r \quad x + 2y = s \quad \text{for } (r, s) \in \mathbb{N}^2. \tag{9.18}$$

Expressed as a polytope problem, it is much easier to see how the
number of vector partitions depends on the actual values of r and s as
we see the geometric implications of the constraints. Figure 9.3 illustrates the three different possibilities.

The matrix defining the problem is given by

$$A = \begin{bmatrix} 1 & 0 & 1 & 1 \\ 0 & 1 & 1 & 2 \end{bmatrix}. \tag{9.19}$$

Then the lattice point enumerator under the parameters (r, s) is the
coefficient

$$[x^{rn} y^{sn}] \frac{1}{(1-x)(1-y)(1-xy)(1-xy^2)}. \tag{9.20}$$

9.5.1 An Exact Solution

This example is simple enough that we can use exact formulas to find the following expression for this value, depending on r and s. We leave the arithmetic details to Exercise 9.4. Working through an exact solution is manageable here because there are only two nontrivial terms. In general, this computation would be more difficult to manage, certainly in higher dimensions. It might be that the simplifications we use to solve the asymptotic formula render the problem more approachable by hand. The solution is summarized in Figure 9.4.

For example, if $r/s \le s \le r$, then $p^{\mathscr{S}}(r,s)$ is the quasi-polynomial:

$$p^{\mathscr{S}}(r,s) = \begin{cases} rs - \frac{s^2}{2} - \frac{r^2}{4} + \frac{r+s}{2} + 1, & \text{if } r \equiv 0 \mod 2 \\ rs - \frac{s^2}{2} - \frac{r^2}{4} + \frac{r+s}{2} + \frac{3}{4}, & \text{if } r \equiv 1 \mod 2 \end{cases}$$

$$= rs - \frac{s^2}{2} - \frac{r^2}{4} + \frac{r+s}{2} + \frac{7}{8} + \frac{(-1)^r}{8}.$$

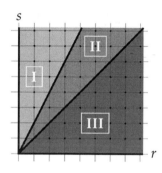

Region		$p^{\mathscr{S}}(r,s)$
I	$2r \le s$	$\frac{r^2}{2} + \frac{3r}{2} + 1$
II	$r \le s \le 2r$	$rs - \frac{r^2}{2} - \frac{s^2}{4} + \frac{r+s}{2} + \frac{7}{8} + \frac{(-1)^s}{8}$
III	$s \le r$	$\frac{s^2}{4} + s + \frac{7}{8} + \frac{(-1)^s}{8}$

FIGURE 9.4

(*top*) The chambers giving the value of $p^{\mathscr{S}}(r,s)$, the number of vector partitions of (r,s) with parts in $\mathscr{S} = \{(1,0),(0,1),(1,1),(1,2)\}$. (*bottom*) An exact solution to the number $p^{\mathscr{S}}(r,s)$ of vector partitions of (r,s) with parts in $\mathscr{S} = \{(1,0),(0,1),(1,1),(1,2)\}$.

9.5.2 An Asymptotic Solution

Next to illustrate the process outlined in this chapter, we express the problem as a diagonal and consider how to apply the various steps. (Even though for this case we know the answer exactly.)

The vector partition generating function can be written as a diagonal of a rational function.

$$\sum_{n\geq 0} p^{\mathscr{S}}(rn, sn)y^n = \Delta^{(r,s)} \frac{1}{(1-x)(1-y)(1-xy)(1-xy^2)}. \tag{9.21}$$

Let us label the factors of H: $H_1 = (1-x), H_2 = (1-y), H_3 = (1-xy), H_4 = (1-xy^2)$. We note that at the critical point $(1,1)$ the intersection cannot be transverse, because it is the intersection of four polynomials (which is larger than the dimension). We note that it is the unique critical point here. Figure 9.5 contains two visualizations of the respective varieties. We see that the series of $F(\mathbf{x})$ that we want converges on the open polydisk $\{(x,y) : |x| < 1, |y| < 1\}$.

9.5.3 The Bases with No Broken Circuits

To make the decomposition, we return to the matrix and label the columns:

$$A = \begin{bmatrix} 1 & 2 & 3 & 4 \\ 1 & 0 & 1 & 1 \\ 0 & 1 & 1 & 2 \end{bmatrix}. \tag{9.22}$$

The matroid $M(A)$ has $E = \{1, 2, 3, 4\}$. The set of bases contains all subsets of size 2. The broken circuits are $\{1, 2\}, \{1, 3\}, \{2, 3\}$. The set of bases that contain no broken circuit is

$$\mathscr{J} = \{\{1, 4\}, \{2, 4\}, \{3, 4\}\}.$$

9.5.4 The Reduction Algorithm

The next step is to find a Nullstellensatz certificate for every broken circuit and use this to reduce the quotient. These examples can be done by inspection. For circuit, $\{1, 2, 3\}$ we notice, $(1 - xy) = y(1 - x) - (1 - y)$ giving rule $H_3 = yH_1 - H_2$. We summarize the rules in Table 9.1.

Next we reduce: For each term, if the indices of the denominator are a set in $\mathscr{A}$, that term is reduced. Otherwise, it contains a broken circuit, and can be reduced using Table 9.1. To avoid duplication, we

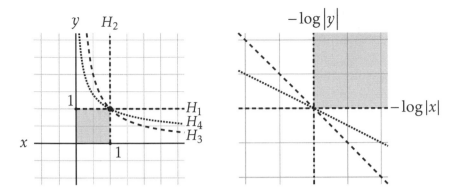

FIGURE 9.5

(left) The positive real points on the variety $\mathcal{V}_{H_i}$, for $i = 1, \ldots, 4$. The shaded area indicates points $(x, y) \in \mathbb{R}^2$ in the domain of convergence of $F(\mathbf{x})$. The intersection at $(1, 1)$ is not transverse. *(right)* The images of these varieties under the -relog map. Remark it is a hyperplane arrangement that contains the origin.

TABLE 9.1

Circuits and Their Rewriting Rules

	Circuit	Rewriting rule
(1)	$\{1, 2, 3\}$	$H_3 = yH_1 - H_2$
(2)	$\{1, 2, 4\}$	$H_4 = H_1 + x(y + 1)H_2$
(3)	$\{1, 3, 4\}$	$H_4 = -yH_1 + (y + 1)H_3$
(4)	$\{2, 3, 4\}$	$H_4 = xyH_2 + H_3$

reduce from largest to smallest, applying a lexicographic order on the circuits. Here is the reduction:

$$\frac{1}{H_1 H_2 H_3 H_4} = \frac{H_4}{H_1 H_2 H_3 H_4^2}$$

$$\overset{(4)}{=} \frac{xyH_2 + H_3}{H_1 H_2 H_3 H_4^2} = \frac{xy}{H_1 H_3 H_4^2} + \frac{1}{H_1 H_2 H_4^2}$$

$$\overset{(3,2)}{=} \frac{xy\left(-yH_1 + (y+1)H_3\right)}{H_1 H_3 H_4^3} + \frac{H_1 + x(y+1)H_2}{H_1 H_2 H_4^3}$$

$$= \frac{-xy^2}{H_3 H_4^3} + \frac{xy(y+1)}{H_1 H_4^3} + \frac{1}{H_2 H_4^3} + \frac{x(y+1)}{H_1 H_4^3}$$

$$= \frac{-xy^2}{H_3 H_4^3} + \frac{x(y+1)^2}{H_1 H_4^3} + \frac{1}{H_2 H_4^3}.$$

TABLE 9.2

The Support Associated to Each Base

Base	Term	Support $x^r y^s$	Region	Contribution
$\{1,4\}$	$(1-x)(1-xy^2)$	$s \leq 2r$	II, III	$s^2/4$
$\{2,4\}$	$(1-y)(1-xy^2)$	$2r \leq s$	I	$r^2/2$
$\{3,4\}$	$(1-xy)(1-xy^2)$	$r \leq s \leq 2r$	II	$(r-s)^2/2$

At this stage, the algorithm terminates, because each term is good: the numerators are non-zero at $(1,1)$, and everything is maximally reduced.

For any of these quotients, the support of the Taylor series will only contain a subset of the possible terms $x^j y^k$. For example, in any given term in the series development of $\frac{1}{(1-y)(1-xy)}$ around the origin, the power of x is always at least the power of y. That is, $\{\mathrm{Supp}\left(\frac{1}{(1-y)(1-xy)}\right) = (j+k,j) \mid j,k \in \mathbb{N}\}$. Each of these rational functions has a restriction on the support. The results are summarized in Table 9.2.

For a fixed (r,s), only a subset of these terms will contribute. Therefore, to determine asymptotic formulas for $p^{\mathscr{S}}(r,s)$, we first determine those summands whose support is in $\{(rn,sn) \mid n \in \mathbb{N}\}$. The decomposition of $F(\mathbf{x})$ is

$$\frac{1}{H} = \frac{x(y+1)^2}{H_1 H_4^3} + \frac{1}{H_2 H_4^3} + \frac{-xy^2}{H_3 H_4^3}. \tag{9.23}$$

9.5.5 Asymptotic Formula

To conclude, we apply Theorem 9.5 tailored to this case. First,

$$\Psi(1,1) := \begin{bmatrix} x\partial_x H_k(x,y) & y\partial_y H_k(x,y) \\ x\partial_x H_\ell(x,y) & y\partial_y H_\ell(x,y) \end{bmatrix}\Bigg|_{(x,y)=(1,1)} = [A_k A_\ell],$$

where A_k and A_ℓ are the k and ℓ-th columns of A. Recall the shorthand: $(r,s)^{(a,b)} := r^a s^b$. Given

$$\frac{G(x,y)}{H_k(x,y)^{m_k} H_\ell(x,y)^{m_\ell}} = \sum_{(r,s)\in\mathbb{N}^2} f(r,s) x^r y^s$$

then,

$$f(r,s) \sim \frac{1}{(m_k-1)!(m_\ell-1)!} \frac{G(1,1)}{\det(\Psi(1,1))} \left((r,s) \times \left(\Psi(1,1)^{-1}\right)\right)^{(m_k-1,m_\ell-1)}$$

for large r, s, provided that (r, s) is in the support of $H_k(x, y)^{m_k} H_\ell(x, y)^{m_\ell}$. We apply this to each of the three terms. Summing the appropriate combinations gives the appropriate estimates for the exact values.

9.6   Discussion

Readers interested in polytope point enumerators will be delighted by the text of Beck and Robins [BR15a]. It is a comprehensive, yet pedagogical treatment of the subject. There are many other references including starting point [Stu95] discussions of computational complexity [DLHTY04]. Computational methods for polytope point enumerators are coded in Latte [JADL]. To directly compute a coefficient numerically is generally unfeasible as the number of terms to manage in the series expansions grow very fast.

The relation between the polynomials indexed by a circuit can be computed using the program singular [DGPS].

A hyperplane arrangement is a finite set $\mathscr{A}$ of hyperplanes in a linear, affine or projective space. Typically, one is interested in the spaces created by removing the hyperplanes. The interplay of geometry and combinatorics is well studied and very elegant. The intersection semilattice determines the key combinatorial invariant of the hyperplane arrangement. The underlying independence results used here is a result of Orlik and Terao [OT92, Theorems 3.43, 3.126 and 5.89].

Leĭnartas' results are in [Leu78], and Raichev [Rai12b] summarized his work in English and has implemented the partial fraction decomposition in Sage.

The example of Section 9.5 has a direct application to algebraic combinatorics. These values can be used to determine Kronecker coefficients. The determination of algebraic quantities motivates much diagonal combinatorics, as we have seen. See [MRS18] for the direct connection of this example to the Kronecker coefficients of representation theory. The work of Vergne and Baldoni [BV18] and [CDW12] address these extractions in that context. The computability of such coefficients has implications in complexity theory.

This chapter barely scratches the surface of arrangement points, much less multiple points. For a comprehensive treatment with explicit examples and formulas, the reader must consult Chapter 10 of [PW13]. Theorem 10.3.7 in particular summarizes the hypotheses and consequences.

9.7 Problems

Exercise 9.1 (Bijective proofs of partition identities). Prove the following identity on formal power series by interpreting each side as generating functions for partitions:

$$1 + \sum_{k \geq 1} \frac{z^k}{(1-z)(1-z^2)\ldots(1-z^k)} = \prod_{k \geq 1} \frac{1}{1-z^k}.$$

$\square$

Exercise 9.2. Determine the generating function for partitions whose parts are all odd numbers. Provide a generating function proof that this class is in bijection with the set of partitions whose parts are distinct. Prove this bijectively. This falls under the spectre of Rogers-Ramanujan identities. $\square$

Exercise 9.3. Consider walks on an integer lattice, taking only $N = (0, 1)$ steps and $E = (1, 0)$. Show that the number of distinct walks from $(0, 0)$ to (a, b) is $\binom{a+b}{a}$. Now consider finite class $\mathscr{P}^{(k,r)}$ of partitions with exactly k summands each at most r. Prove that as $z \to 1$ the generating function reduces to $\binom{k+r-1}{k}$. $\square$

Exercise 9.4. Using direct coefficient extraction and series manipulation to prove the correctness of Figure 9.4. $\square$

Exercise 9.5. Show that the order of the nonconstant part of the the subexponential growth of an integer partition is always determined by the behaviour of the generating function at one. To what extent is the analogue true in higher dimensions?

Reference: [Kou08]

$\square$

Exercise 9.6. What can you say about the expected value of the r-th smallest part in a partition of n?

Reference: [GMPR14]

$\square$

Exercise 9.7. Find the asymptotic number of vector partitions with parts from $\mathscr{S} = \{(1,0),(0,1),(2,0),(2,3)\}$. Determine the number of chambers in the solution.   ❏

Exercise 9.8. Prove that the lattice point enumerator of $n\mathbb{P}_{d-1}$ is $\binom{d+n-1}{d}$.   ❏

Bibliography

[AB12]   Boris Adamczewski and Jason P. Bell. On vanishing co-
           efficients of algebraic power series over fields of positive
           characteristic. *Invent. Math.*, 187(2):343–393, 2012.

[AB13]   Boris Adamczewski and Jason P. Bell. Diagonalization
           and rationalization of algebraic Laurent series. *Ann. Sci.
           Éc. Norm. Supér. (4)*, 46(6):963–1004, 2013.

[Anı71]   A. V. Anısımov. The group languages. *Kibernetika (Kiev)*,
           (4):18–24, 1971.

[JADL]   Velleda Baldoni, Nicole Berline, Jesús A De Loera, Bran-
           don Dutra, Matthias Köppe, Stanislav Moreinis, Gre-
           gory Pinto, Michele Vergne, and Jianqiu Wu. A user's
           guide for latte integrale v1. 7.2. *Optimization*, 22(2),
           2014.

[BV18]   Velleda Baldoni and Michele Vergne. Computation of
           dilated Kronecker coefficients. *J. Symbolic Comput.*,
           84:113–146, 2018. With an appendix by M. Walter.

[BD15]   Cyril Banderier and Michael Drmota. Formulae and
           asymptotics for coefficients of algebraic functions. *Com-
           bin. Probab. Comput.*, 24(1):1–53, 2015.

[BLFR01]   Elena Barcucci, Alberto Del Lungo, Andrea Frosini, and
           Simone Rinaldi. A technology for reverse-engineering a
           combinatorial problem from a rational generating func-
           tion. *Adv. in Appl. Math.*, 26(2):129–153, 2001.

[BR15a]   Matthias Beck and Sinai Robins. *Computing the contin-
           uous discretely*. Undergraduate Texts in Mathematics.
           Springer, New York, second edition, 2015. Integer-point
           enumeration in polyhedra, with illustrations by David
           Austin.

213

[BM18]   Jason P. Bell and Marni Mishna.   On the complex-
            ity of the cogrowth sequence.   https://arxiv.org/abs/
            1805.08118[Math.CO], 2018.

[BC17]   Jason P. Bell and Shaoshi Chen. Power series with co-
            efficients from a finite set. *J. Combin. Theory Ser. A*,
            151:241–253, 2017.

[Ben74]   Edward A. Bender. Asymptotic methods in enumera-
            tion. *SIAM Rev.*, 16:485–515, 1974.

[BLL98]   F. Bergeron, G. Labelle, and P. Leroux. *Combinatorial
            species and tree-like structures*, volume 67 of *Encyclopedia
            of Mathematics and its Applications*. Cambridge Univer-
            sity Press, Cambridge, 1998. Translated from the 1994
            French original by Margaret Readdy, with a foreword by
            Gian-Carlo Rota.

[BR11]   Jean Berstel and Christophe Reutenauer. *Noncommuta-
            tive rational series with applications*, volume 137 of *Ency-
            clopedia of Mathematics and its Applications*. Cambridge
            University Press, Cambridge, 2011.

[BM93]   Andrea Bertozzi and James McKenna.   Multidimen-
            sional residues, generating functions, and their applica-
            tion to queueing networks. *SIAM Rev.*, 35(2):239–268,
            1993.

[Bia92]   Philippe Biane. Minuscule weights and random walks
            on lattices. In *Quantum probability & related topics*, QP-
            PQ, VII, pages 51–65. World Sci. Publ., River Edge, NJ,
            1992.

[Bia93]   Philippe Biane. Estimation asymptotique des multi-
            plicités dans les puissances tensorielles d'un *g*-module.
            C. R. Acad. Sci. Paris Sér. I Math., 316(8):849–852, 1993.

[BFKV07]   Manuel Bodirsky, Eric Fusy, Mihyun Kang, and Stefan
            Vigerske. An unbiased pointing operator for unlabeled
            structures, with applications to counting and sampling.
            In *Proceedings of the Eighteenth Annual ACM-SIAM Sym-
            posium on Discrete Algorithms*, pages 356–365. ACM,
            New York, 2007.

[BFKV11]   Manuel Bodirsky, Éric Fusy, Mihyun Kang, and Stefan Vigerske. Boltzmann samplers, Pólya theory, and cycle pointing. *SIAM J. Comput.*, 40(3):721–769, 2011.

[BBC⁺13]   A. Bostan, S. Boukraa, G. Christol, S. Hassani, and J.-M. Maillard. Ising n-fold integrals as diagonals of rational functions and integrality of series expansions. *J. Phys. A*, 46(18):185202, 44, 2013.

[BBMKM16] Alin Bostan, Mireille Bousquet-Mélou, Manuel Kauers, and Stephen Melczer. On 3-dimensional lattice walks confined to the positive octant. *Ann. Comb.*, 20(4):661–704, 2016.

[BBMM]   Alin Bostan, Mireille Bousquet-Mélou, and Stephen Melczer. Counting walks with large steps in an orthant. 2018. ⟨hal-01802706⟩

[BCG⁺17]   Alin Bostan, Frédéric Chyzak, Marc Giusti, Romain Lebreton, Grégoire Lecerf, Bruno Salvy, and Éric Schost. *Algorithmes Efficaces en Calcul Formel*. Frédéric Chyzak (auto-édit.), Palaiseau, September 2017. 686 pages.

[BDS17]   Alin Bostan, Louis Dumont, and Bruno Salvy. Algebraic diagonals and walks: algorithms, bounds, complexity. *J. Symbolic Comput.*, 83:68–92, 2017.

[BLS17]   Alin Bostan, Pierre Lairez, and Bruno Salvy. Multiple binomial sums. *J. Symbolic Comput.*, 80(part 2):351–386, 2017.

[BR15b]   Alin Bostan and Kilian Raschel. Compter les excursions sur un échiquier. *Pour la science*, (449):40–46, 2015.

[BRS14]   Alin Bostan, Kilian Raschel, and Bruno Salvy. Non-D-finite excursions in the quarter plane. *J. Comb. Theory, Ser. A*, 121(0):45–63, 2014.

[BM02]   Mireille Bousquet-Mélou. Counting walks in the quarter plane. In *Mathematics and computer science, II (Versailles, 2002)*, Trends Math., pages 49–67. Birkhäuser, 2002.

[BM16]   Mireille Bousquet-Mélou. Square lattice walks avoiding a quadrant. *J. Combin. Theory Ser. A*, 144:37–79, 2016.

[BMM10]   Mireille Bousquet-Mélou and Marni Mishna. Walks with small steps in the quarter plane. In *Algorithmic Probability and Combinatorics*, volume 520 of *Contemp. Math.*, pages 1–40. Amer. Math. Soc., 2010.

[BMP00]   Mireille Bousquet-Mélou and Marko Petkovšek. Linear recurrences with constant coefficients: the multivariate case. *Discrete Mathematics*, 225(1-3):51–75, 2000.

[BMR03]   Mireille Bousquet-Mélou and Andrew Rechnitzer. The site-perimeter of bargraphs. *Adv. in Appl. Math.*, 31(1):86–112, 2003.

[BGPP10]   Andrew Bressler, Torin Greenwood, Robin Pemantle, and Marko Petkovšek. Quantum random walk on the integer lattice: examples and phenomena. In *Algorithmic probability and combinatorics*, volume 520 of *Contemp. Math.*, pages 41–60. Amer. Math. Soc., Providence, RI, 2010.

[Bud]   Timothy Budd. Winding of simple walks on the square lattice. https://arxiv.org/abs/1709.04042, 2017.

[Bos]   Alin Bostan. Calcul formel pour la combinatoire des marches - Habilitation a Diriger des Recherches, Universite Paris, 2017.

[CDNS11]   John S. Caughman, Charles L. Dunn, Nancy Ann Neudauer, and Colin L. Starr. Counting lattice chains and delannoy paths in higher dimensions. *Discrete Mathematics*, 311(16):1803–1812, 2011.

[CDW12]   Matthias Christandl, Brent Doran, and Michael Walter. Computing multiplicities of Lie group representations. In *2012 IEEE 53rd Annual Symposium on Foundations of Computer Science—FOCS 2012*, pages 639–648. IEEE Computer Soc., Los Alamitos, CA, 2012.

[Chr90]   Gilles Christol. Globally bounded solutions of differential equations. In *Analytic number theory (Tokyo, 1988)*, volume 1434 of *Lecture Notes in Math.*, pages 45–64. Springer, Berlin Heidelberg, 1990.

[CC85]   D. V. Chudnovsky and G. V. Chudnovsky. Applications of Padé approximations to Diophantine inequalities in values of *G*-functions. In *Number theory* (New York, 1983–84), volume 1135 of *Lecture Notes in Math.*, pages 9–51. Springer, Berlin, 1985.

[CY18]   Frédéric Chyzak and Karen Yeats. Bijections between Łukasiewicz walks and generalized tandem walks. https://arxiv.org/abs/1810.04117, 2018.

[CMMR17] J. Courtiel, S. Melczer, M. Mishna, and K. Raschel. Weighted lattice walks and universality classes. *J. Combin. Theory Ser. A*, 152:255–302, 2017.

[DLHTY04] Jesús A. De Loera, Raymond Hemmecke, Jeremiah Tauzer, and Ruriko Yoshida. Effective lattice point counting in rational convex polytopes. *J. Symbolic Comput.*, 38(4):1273–1302, 2004.

[Del94]   Maylis Delest. Formal calculus and enumerative combinatorics. In *Computational support for discrete mathematics (Piscataway, NJ, 1992)*, volume 15 of *DIMACS Ser. Discrete Math. Theoret. Comput. Sci.*, pages 335–345. Amer. Math. Soc., Providence, RI, 1994.

[DL87]   J. Denef and L. Lipshitz. Algebraic power series and diagonals. *J. Number Theory*, 26(1):46–67, 1987.

[DGPR10]   Li Dong, Zhicheng Gao, Daniel Panario, and Bruce Richmond. Asymptotics of smallest component sizes in decomposable combinatorial structures of alg-log type. *Discrete Math. Theor. Comput. Sci.*, 12(2):197–222, 2010.

[DHRS18]   Thomas Dreyfus, Charlotte Hardouin, Julien Roques, and Michael F. Singer. On the nature of the generating series of walks in the quarter plane. *Invent. Math.*, 213(1):139–203, 2018.

[Drm94]   Michael Drmota. A bivariate asymptotic expansion of coefficients of powers of generating functions. *European J. Combin.*, 15(2):139–152, 1994.

[Dun85]   M. J. Dunwoody. The accessibility of finitely presented groups. *Invent. Math.*, 81(3):449–457, 1985.

[ERJvRW14] Murray Elder, Andrew Rechnitzer, Esaias J. Janse van Rensburg, and Thomas Wong. The cogrowth series for BS(N, N) is D-finite. *Internat. J. Algebra Comput.*, 24(2):171–187, 2014.

[Fat06]   P. Fatou. Séries trigonométriques et séries de Taylor. *Acta Math.*, 30(1):335–400, 1906.

[Fei10]   Thomas Feierl. Asymptotics for walks in a Weyl chamber of type B (extended abstract). In *21st International Meeting on Probabilistic, Combinatorial, and Asymptotic Methods in the Analysis of Algorithms (AofA'10)*, Discrete Math. Theor. Comput. Sci. Proc., AM, pages 175–188. Assoc. Discrete Math. Theor. Comput. Sci., Nancy, 2010.

[Fla87]   Philippe Flajolet. Analytic models and ambiguity of context-free languages. *Theoret. Comput. Sci.*, 49(2-3):283–309, 1987. Twelfth international colloquium on automata, languages and programming (Nafplion, 1985).

[FGOR93]   Philippe Flajolet, Zhicheng Gao, Andrew Odlyzko, and Bruce Richmond. The distribution of heights of binary trees and other simple trees. *Combin. Probab. Comput.*, 2(2):145–156, 1993.

[FGT92]   Philippe Flajolet, Danièle Gardy, and Loÿs Thimonier. Birthday paradox, coupon collectors, caching algorithms and self-organizing search. *Discrete Appl. Math.*, 39(3):207–229, 1992.

[FGS06]   Philippe Flajolet, Stefan Gerhold, and Bruno Salvy. On the non-holonomic character of logarithms, powers, and the nth prime function. *Electron. J. Combin.*, 11(2):Article 2, 16, 2004/06.

[FO90]   Philippe Flajolet and Andrew Odlyzko. Singularity analysis of generating functions. *SIAM J. Discrete Math.*, 3(2):216–240, 1990.

[FS09]   Philippe Flajolet and Robert Sedgewick. *Analytic combinatorics*. Cambridge University Press, Cambridge, 2009.

[GMPR14]   Zhicheng Gao, Conrado Martinez, Daniel Panario, and Bruce Richmond. The rth smallest part size of a random integer partition. *Integers*, 14A:Paper No. A3, 8, 2014.

[GR92]   Zhicheng Gao and L. Bruce Richmond. Central and local limit theorems applied to asymptotic enumeration. IV. Multivariate generating functions. *J. Comput. Appl. Math.*, 41(1-2):177–186, 1992. Asymptotic methods in analysis and combinatorics.

[GP]   Scott Garrabrant and Igor Pak. Counting with irrational tiles. http://arxiv.org/abs/1407.8222.

[GP17]   Scott Garrabrant and Igor Pak. Words in linear groups, random walks, automata and P-recursiveness. *J. Comb. Algebra*, 1(2):127–144, 2017.

[Ges81]   Ira Gessel. Two theorems of rational power series. *Utilitas Math.*, 19:247–254, 1981.

[Ges90]   Ira Gessel. Symmetric functions and P-recursiveness. *J. Combin. Theory Ser. A*, 53(2):257–285, 1990.

[Ges03]   Ira Gessel. Rational functions with nonnegative integer coefficients. Slides from a talk at the 50th Seminaire Lotharingien de Combinatoire, 2003.

[GZ92]   Ira Gessel and Doron Zeilberger. Random walk in a Weyl chamber. *Proc. Amer. Math. Soc.*, 115(1):27–31, 1992.

[Gra06]   David J. Grabiner. Asymptotics for the distributions of subtableaux in Young and up-down tableaux. *Electron. J. Combin.*, 11(2):Research Paper 29, 22, 2004/06.

[Gra02]   David J. Grabiner. Random walk in an alcove of an affine Weyl group, and non-colliding random walks on an interval. *J. Combin. Theory Ser. A*, 97(2):285–306, 2002.

[Gre83]   D.H. Greene. *Labelled Formal Languages and Their Uses*. PhD thesis, 1983.

[DGPS]   Gert-Martin Greuel, Gerhard Pfister, and Hans Schönemann. Singular: a computer algebra system for polynomial computations. *ACM Communications in Computer Algebra*, 42(3):180–181, 2009.

[Gut00]   A. J. Guttmann. Indicators of solvability for lattice models. *Discrete Math.*, 217(1-3):167–189, 2000. Formal power series and algebraic combinatorics (Vienna, 1997).

[Hŏ90]   L. Hörmander. *The analysis of linear partial differential operators. I*, volume 256 of *Grundlehren der Mathematischen Wissenschaften*. Springer-Verlag, Berlin, second edition, 1990. Distribution theory and Fourier analysis.

[Hum92]   James E. Humphreys. *Reflection groups and Coxeter groups*, volume 29. Cambridge University Press, 1992.

[Kes59]   Harry Kesten. Symmetric random walks on groups. *Trans. Amer. Math. Soc.*, 92:336–354, 1959.

[Kou08]   Christoph Koutschan. Regular languages and their generating functions: the inverse problem. *Theoret. Comput. Sci.*, 391(1-2):65–74, 2008.

[Kra07]   Christian Krattenthaler. Asymptotics for random walks in alcoves of affine Weyl groups. *Sém. Lothar. Combin.*, 52:Art. B52i, 72, 2004/07.

[Kra07]   Christian Krattenthaler. Asymptotics for random walks in alcoves of affine weyl groups. *Séminaire Lotharingien de Combinatoire*, 52:B52i, 2007.

[Leu78]   E. K. Leunartas. Factorization of rational functions of several variables into partial fractions. *Izv. Vyssh. Uchebn. Zaved. Mat.*, (10(197)):47–51, 1978.

[Lip88]   L. Lipshitz. The diagonal of a *D*-finite power series is *D*-finite. *J. Algebra*, 113(2):373–378, 1988.

[LvdP90]   Leonard Lipshitz and Alfred J. van der Poorten. Rational functions, diagonals, automata and arithmetic. In *Number theory (Banff, AB, 1988)*, pages 339–358. de Gruyter, Berlin, 1990.

[Lot83]   M. Lothaire. *Combinatorics on words*, volume 17 of *En-cyclopedia of Mathematics and its Applications*. Addison-Wesley Publishing Co., Reading, Mass., 1983. A collective work by Dominique Perrin, Jean Berstel, Christian Choffrut, Robert Cori, Dominique Foata, Jean Eric Pin, Guiseppe Pirillo, Christophe Reutenauer, Marcel-P. Schützenberger, Jacques Sakarovitch, and Imre Simon, With a foreword by Roger Lyndon, Edited and with a preface by Perrin.

[Mel17]   Stephen Melczer. *Analytic Combinatorics in Several Variables: Effective Asymptotics and Lattice Path Enumeration*. PhD thesis, University of Waterloo and ENS Lyon, 2017. Thesis (PhD).

[MM16]   Stephen Melczer and Marni Mishna. Asymptotic lattice path enumeration using diagonals. *Algorithmica*, 75(4):782–811, 2016.

[MS16]   Stephen Melczer and Bruno Salvy. Symbolic-numeric tools for analytic combinatorics in several variables. In *Proceedings of the 2016 ACM International Symposium on Symbolic and Algebraic Computation*, pages 333–340. ACM, New York, 2016.

[MW18]   Stephen Melczer and Mark C. Wilson. Higher dimensional lattice walks: Connecting combinatorial and analytic behavior. https://arxiv.org/abs/1810.06170, 2018.

[Mil35]   E. W. Miller. On the singularities of an analytic function. *Bull. Amer. Math. Soc.*, 41(8):561–565, 1935.

[Mis19]   Marni Mishna. On standard young tableaux of bounded height. In *Recent Trends in Algebraic Combinatorics*, pages 281–303. Springer, 2019.

[MR09]   Marni Mishna and Andrew Rechnitzer. Two non-holonomic lattice walks in the quarter plane. *Theoret. Comput. Sci.*, 410(38-40):3616–3630, 2009.

[MRS18]   Marni Mishna, Mercedes Rosas, and Sheila Sundaram. An elementary approach to the quasipolynomiality of the kronecker coefficients. https://arxiv.org/abs/1811.10015, 2018.

[MS19]   Marni Mishna and Samuel Simon. The asymptotics of reflectable weighted random walks in arbitrary dimension. In *Proceedings of EUROCOMB 2019*. Acta Math. Univ. Comenianae, 2019.

[MS83]   David E. Muller and Paul E. Schupp. Groups, the theory of ends, and context-free languages. *J. Comput. System Sci.*, 26(3):295–310, 1983.

[Oga14]   Hiroshi Ogawara.   Differential transcendency of a formal laurent series satisfying a rational linear q-difference equation. *Funkcialaj Ekvacioj*, 57:477–488, 2014.

[OT92]   Peter Orlik and Hiroaki Terao. *Arrangements of hyperplanes*, volume 300 of *Grundlehren der Mathematischen Wissenschaften [Fundamental Principles of Mathematical Sciences]*. Springer-Verlag, Berlin, 1992.

[Pak18]   Igor Pak. Complexity problems in enumerative combinatorics. In *Proc. ICM Rio de Janeiro*, volume 3, pages 3139–3166, 2018.   https://arxiv.org/abs/1803.06636, 2018.

[PW08]   Robin Pemantle and Mark C Wilson. Twenty combinatorial examples of asymptotics derived from multivariate generating functions. *Siam Review*, 50(2):199–272, 2008.

[PW13]   Robin Pemantle and Mark Wilson. *Analytic combinatorics in several variables*, volume 140 of *Cambridge Studies in Advanced Mathematics*.   Cambridge University Press, Cambridge, 2013.

[PSS12]   Carine Pivoteau, Bruno Salvy, and Michèle Soria. Algorithms for combinatorial structures: well-founded systems and Newton iterations. *J. Combin. Theory Ser. A*, 119(8):1711–1773, 2012.

[Rai12a]   A. Raichev. Leinartas's partial fraction decomposition. https://arxiv.org/abs/1206.4740, 2018.

[Rai12b]   A. Raichev. Leinartas's partial fraction decomposition. Technical Report CDMTCS-421, University of Auckland, Centre for Discrete Mathematics and Theoretical Computer Science Research Reports, 2012.

[RW11]   Alexander Raichev and Mark C. Wilson. Asymptotics of coefficients of multivariate generating functions: improvements for multiple points. *Online J. Anal. Comb.*, (6):21, 2011.

[RT]   Kilian Raschel and Amelie Trotignon. On walks avoiding a quadrant. *Electronic Journal of Combinatorics*, 26:P3.31.

[Rec06a]   Andrew Rechnitzer. Haruspicy 2: the anisotropic generating function of self-avoiding polygons is not D-finite. *J. Combin. Theory Ser. A*, 113(3):520–546, 2006.

[Rec06b]   Andrew Rechnitzer. Haruspicy 3: the anisotropic generating function of directed bond-animals is not D-finite. *J. Combin. Theory Ser. A*, 113(6):1031–1049, 2006.

[Rec09]   Andrew Rechnitzer. The anisotropic generating function of self-avoiding polygons is not D-finite. In *Polygons, polyominoes and polycubes*, volume 775 of *Lecture Notes in Phys.*, pages 93–115. Springer, Dordrecht, 2009.

[Rem95]   Reinhold Remmert. *Funktionentheorie 2*. Springer-Lehrbuch,. Springer Berlin Heidelberg, Berlin, Heidelberg, zweite, korrigierte auflage, edition, 1995.

[Rub89]   Lee A. Rubel. A survey of transcendentally transcendental functions. *Amer. Math. Monthly*, 96(9):777–788, 1989.

[SZ94]   Bruno Salvy and Paul Zimmermann. Gfun: a Maple package for the manipulation of generating and holonomic functions in one variable. *ACM Transactions on Mathematical Software*, 20(2):163–177, 1994.

[SS17]   Reinhard Schäfke and Michael F Singer. Consistent systems of linear differential and difference equations. https://arxiv.org/abs/1605.02616, 2016.

[Sta80]   Richard Stanley. Differentiably finite power series. *European J. Combin.*, 1(2):175–188, 1980.

[Sta99]   Richard Stanley. *Enumerative combinatorics*, volume 2. Cambridge University Press, 1999.

[Sta15]   Richard P. Stanley. *Catalan numbers*. Cambridge University Press, New York, 2015.

[Stu95]   Bernd Sturmfels. On vector partition functions. *Combin.Theory Ser A*, 72(2), 1995.

[Tat11]   Tatsuya Tate.   Problems on asymptotic analysis over convex polytopes.   In *Geometry and quantization*, volume 19 of *Trav. Math.*, pages 65–96. Univ. Luxemb., Luxembourg, 2011.

[TZ04]   Tatsuya Tate and Steve Zelditch.   Lattice path combinatorics and asymptotics of multiplicities of weights in tensor powers. *J. Funct. Anal.*, 217(2):402–447, 2004.

[vdPS96]   A. J. van der Poorten and I. E. Shparlinski. On linear recurrence sequences with polynomial coefficients. *Glasgow Math. J.*, 38(2):147–155, 1996.

[Was65]   W. Wasow. *Asymptotic expansions for ordinary differential equations*. Interscience Publishers, 1965.

[WS89]   Christopher F. Woodcock and Habib Sharif. On the transcendence of certain series. *J. Algebra*, 121(2):364–369, 1989.

[WC13]   Xiao Li Wu and Shao Shi Chen. A note on the diagonal theorem for bivariate rational formal power series. *Acta Math. Sinica (Chin. Ser.)*, 56(2):203–210, 2013.

[Zei83]   Doron Zeilberger. Andre's reflection proof generalized to the many-candidate ballot problem. *Discrete Mathematics*, 44(3):325–326, 1983.

[Zei90]   D. Zeilberger. A holonomic systems approach to special functions identities. *J. Comput. Appl. Math.*, 32(3):321–368, 1990.

Glossary

algebraic function: A function that satisfies a polynomial equation.

analytic at a point: The property to have a convergent series expansion at a point, with a non-zero radius of convergence.

combinatorial class: A set equipped with a size function satisfying the condition that the number of points of a given size is finite.

combinatorial series: A series with all its coefficients in $\mathbb{N}$.

combinatorial sum: An operation on two sets equivalent to the disjoint union.

critical point: A solution to the critical point equations.

derived class: A subclass of an algebraic combinatorial class defined by a set of possible values of an inherited parameter.

D-finite function: A function that satisfies a linear differential equation with polynomial coefficients.

differentiably algebraic: The property to satisfy a (possibly nonlinear) differential equation with polynomial coefficients.

domain of convergence: The interior of the set of points where the series converges.

entire function: A function that is analytic over all of $\mathbb{C}$.

finite minimal: A property of a critical point that indicates there are a finite number of critical points on its torus.

holomorphic at a point: The property to be differentiable in a neighbourhood of that point.

hypertranscendental: The property of a function to not be differentiably algebraic.

language over an alphabet: A subset of all possible sequences of symbols from the alphabet.

manifold: A collection of points with the property that near each point it resembles Euclidean space ($\mathbb{R}^d$).

minimal point of rational function: A point in the intersection of the boundary of convergence of a series expansion and the singular variety.

multiple point: A point where a variety locally decomposes as a union of smooth varieties.

$\mathbb{N}$-rational: The property of a function to be an ordinary generating function of an S-regular combinatorial class.

$\mathbb{N}$-algebraic: The property of a function to be the ordinary generating function of an algebraic combinatorial class.

ordinary generating function (OGF): The formal power that encodes a counting sequence in its coefficients.

partition: A weakly decreasing sequence of positive integers.

quasi-inverse: A map that sends F to $\frac{1}{1-F}$.

simple walk: A sequence of elements of $\mathscr{S}$ with

$$\mathscr{S} = \{\pm e_j \mid j = 1\dots d\} = \{(\pm 1, 0, \dots, 0), (0, \pm 1, 0, \dots,), \cdots, (0, \cdots, 0, \pm 1)\}.$$

singularity: A point where a function is not analytic (equivalently, not holomorphic).

strictly minimal: The property of a critical point to be the only critical point on its associated torus.

torus associated to a point: The set of points in complex space that have the same modulus for each component.

transcendental: The property to not be algebraic.

transverse multiple point: A point in the intersection of more than one variety with the property that the normals to the varieties are linearly independent. Intuitively, the intersection is robust to small perturbations of the curves.

variety: An algebraic set of points given by the vanishing of a polynomial.

Index

$C_G^{\mathscr{S}}(z)$, 131
$C_\chi(x)$, 37
$\Gamma(x)$, 158
Ω-restricted trees, 22
$\Phi_\rho(n)$, 177
$\mathbb{N}$-rational functions, 76

absolutely convergent, 82
admissible operations, 8
algebraic class, 24
algebraic function, 113
analytic function, 83
aperiodic, 103
arrangements points, 192
atomic class, 14
augmented log normal matrix, 188

balanced classes, 63
balanced words, 151
binomial sum, 58

Catalan numbers, 29
catalytic variables, 49
Cauchy integral formula, 94
    multivariable, 162
Cauchy's Residue Theorem, 94
Cayley graph, 130
central diagonal, 56
central limit theorem, 172
central weights, 184
co-growth series, 131
combinatorial bijection, 7

combinatorial class, 4
    algebraic, 24
    derived, 56, 63
    S-regular, 18
    transcendental, 55, 63
combinatorial series, 160
combinatorial sum, 9
compositions, 31
constant term extraction, 56
context-free language, 26
counting sequence, 4
critical point, 147
    arrangement point, 192
    contributing, 148
    criteria, 181
    isolated, 148
    minimal, 141
    multiple, 177
    smooth, 156
    strictly minimal, 142
    visualization, 180

D-finite, 58, 115
    closure properties, 116
    criteria, 117, 132
    lattice path models, 127
Daffodil lemma, 103
Delannoy numbers, 63
derivation tree, 25
diagonal, 56, 117
differentiably algebraic, 121
differentiably finite, 115
domain of convergence, 141

dominant singularities, 87
Dyck paths, 8, 29

elementary basis vector, 6
entire function, 83, 93
epsilon class, 14
excursions, 7, 50

Ferrers diagram, 192
finite reflection groups, 70
finite state machine, 19
first return decomposition,
    16
formal power series, 11
Fourier-Laplace integral,
    159

G-function, 120
gamma function, 158
generalized binomial, 186
generating function
    bivariate, 36
    cumulative, 37
    exponential (EGF), 12
    multivariate, 41
    ordinary (OGF), 12
    probability, 159
    probability (PGF), 12
    section, 37
Gessel walks, 33

Hadamard product, 117
holomorphic function, 83

integer composition, 44
integer partition, 192
inventory polynomial of $\mathscr{S}$,
    62

kernel equation, 126
kernel method, 50
Kreweras walks, 33

lattice path models, 6
    classification, 124
    Delannoy, 63, 150, 172
    Dyck paths, 8, 49, 66
    Gessel, 33
    Kreweras, 33
    simple, 68, 164
    simple walk, 6
    simple weighted walks,
        184
    tandem, 73, 182, 188
lattice point enumerator, 196
Laurent expansion, 85
Laurent polynomials, 58
Laurent series, 58

Mahler equation, 118, 124
manifold, stratification, 178
matroid, 201
meromorphic, 85
minimal point, 141
moment, factorial, 38
Motzkin paths, 6
multiple point, 177, 179
    contribution, 187
    criteria, 182
    order, 179
    transversal, 180

natural boundary, 84, 136
non-plane trees, 32

OEIS
    A000079, 8
    A000108, 29
    A001006, 8
    A001850, 63
    A002894, 62
    A005717, 64
    A094423, 113
    A135404, 33

P-recursive sequence, 58
parameter, 36
    additive, 46
    inherited, 42
    multidimensional, 41
plane tree, 21
polydisk, 141
polytope, 196
polytope point enumerator, 192
pumping lemma
    context-free languages, 27

rational function, 112
reflectable step set, 71
reflection principle, 65
regular specification, 18
relog, 146
residue, 93

saddle-point approximation, 158
second principle of coefficient
    asymptotics, 99
sequence
    period, 103
    support, 103
sequence operator, 10
series
    combinatorial, 56, 88, 142
    iterated Laurent series, 59

Laurent, 58, 85
    period, 103
    Puiseaux, 86
    section, 108
    span, 103
    support, 103
simple walks, 6, 164
singular variety, 141
singularity
    dominant, 87
    isolated, 83
    pole, 85
smooth point, contribution, 167,
    171
Stirling's approximation, 157
stratum, 178
strictly minimal, 142

torus, 141
transversal intersection,
    180
trees, 21
trinomial, 149

vector partition, 192

Weierstrass M-test, 82
Weyl group, 71
word problem, 130

Printed and bound by CPI Group (UK) Ltd, Croydon, CR0 4YY

17/10/2024

01775655-0014